Discerning Experts

Discerning Experts

The Practices of Scientific Assessment for Environmental Policy

MICHAEL OPPENHEIMER,
NAOMI ORESKES, DALE JAMIESON,
KEYNYN BRYSSE, JESSICA O'REILLY,
MATTHEW SHINDELL, AND
MILENA WAZECK

University of Chicago Press

CHICAGO AND LONDON

The University of Chicago Press, Chicago 60637
The University of Chicago Press, Ltd., London

Published 2019
Printed in the United States of America

28 27 26 25 24 23 22 21 20 19   1 2 3 4 5

ISBN-13: 978-0-226-60196-0 (cloth)
ISBN-13: 978-0-226-60201-1 (paper)
ISBN-13: 978-0-226-60215-8 (e-book)
DOI: https://doi.org/10.7208/chicago/9780226602158.001.0001

Library of Congress Cataloging-in-Publication Data

Names: Oppenheimer, Michael, author. | Oreskes, Naomi, author. | Jamieson, Dale, author. | Brysse, Keynyn, author. | O'Reilly, Jessica, 1978– author. | Shindell, Matthew, author. | Wazeck, Milena, author.
Title: Discerning experts : the practices of scientific assessment for environmental policy / Michael Oppenheimer, Naomi Oreskes, Dale Jamieson, Keynyn Brysse, Jessica O'Reilly, Matthew Shindell, and Milena Wazeck.
Description: Chicago ; London : University of Chicago Press, 2019. | Includes bibliographical references and index.
Identifiers: LCCN 2018021455 | ISBN 9780226601960 (cloth : alk. paper) | ISBN 9780226602011 (pbk. : alk. paper) | ISBN 9780226602158 (e-book)
Subjects: LCSH: Environmental sciences—Research—Evaluation. | Environmental sciences—Research—Evaluation—Case studies. | Acid rain—Research—United States—Evaluation. | National Acid Precipitation Assessment Program (U.S.) | Ozone layer depletion—Research—Evaluation. | Sea level—Research—Antarctica—Evaluation. | Environmental policy—Research—United States. | Research—Evaluation. | Expertise.
Classification: LCC GE70 .O66 2019 | DDC 363.7/0561—dc23
LC record available at https://lccn.loc.gov/2018021455

♾ This paper meets the requirements of ANSI/NISO Z39.48-1992 (Permanence of Paper).

We dedicate this book to
Stephen H. Schneider
(1945–2010)
friend, colleague, inspiration
and
Bert Bolin
(1925–2007)
for his foresight and steady leadership

Contents

Preface

This book, and the project it embodies, originated with a chance encounter between Naomi Oreskes and Michael Oppenheimer at the annual meeting of the American Geophysical Union at the end of 2006. Oreskes has a long involvement in questions about the nature of scientific knowledge, from the perspective of both a geologist and a science studies scholar. Oppenheimer is a climate scientist with a long record of participation in ozone and climate assessments, including the then-ongoing *Fourth Assessment Report of the Intergovernmental Panel on Climate Change*, and he had become intrigued by the processes by which experts decide on matters of fact and uncertainty and how expert judgment has sometimes gone awry. Both were curious about the inner workings of assessments and how scientists arrived at their conclusions. Oreskes and Oppenheimer engaged the interest of Dale Jamieson, a philosopher of science who had studied the National Acid Precipitation Assessment Program. Together the three designed a project to investigate how assessments operate in practice—how assessors come to their views and make the requisite judgments about the state of scientific knowledge.

International expert assessments have become an approach favored by governments for obtaining advice on the science, economics, and policy options available to confront large-scale environmental problems. Assessments that emerged in the 1970s altered both the policy landscape and the practice of environmental science. Understanding how assessments operate requires reconstructing the deliberations of expert participants, who reach judgments about what is known and what is uncertain in the scientific and political context of their times.

Previous work, such as studies carried out by the Global Environmental Assessment Project,[1] focused on aspects of environmental assessments that render them effective in the policy domain. In contrast, our study is about scientific experts and the processes by which questions are framed and judgments are rendered within assessments. The focus of our project is largely internal to the institutions performing assessments and to the deliberations among authors.

Our initial work centered on a series of assessments of the behavior of the West Antarctic Ice Sheet (WAIS), initiated by the project's first postdoctoral fellow, Jessica O'Reilly, an ethnographer experienced in studying Antarctic scientists. Soon the project, dubbed Assessing Assessments, received support from the National Science Foundation. Over time, we were joined by three additional highly talented postdoctoral fellows: Keynyn Brysse, a historian of paleontology who has examined major controversies, including the mass extinction debates and the reclassifications of the Burgess Shale organisms; Milena Wazeck, a historian of science and expert on Albert Einstein and controversies over the theory of relativity; and Matthew Shindell, a historian of science with a focus on earth and planetary sciences, and the biographer of American chemist Harold Urey. Their participation in the project allowed us to expand our focus to two additional sets of assessments: the National Acid Precipitation Assessment Program (NAPAP) and the multiple national and international assessments of ozone depletion. Our methods consisted of interviews of participants and analyses of archival records.

We focused on these particular assessments for several reasons. First, these were all assessments of earth and environmental science (our primary expertise and interest) that played a significant role in crucial public policy debates of the past 40 years. The various assessments of ozone (chapter 3), for instance, contributed directly to the development of international law, including the Montreal Protocol on Substances That Deplete the Ozone Layer, which has helped to protect the very existence of life on earth. Conversely, some argue that NAPAP (chapter 2) contributed to delay in US policy on acid rain. While action on climate change to date has not been commensurate with the scale of the problem, there is no question that the WAIS assessments we studied (chapter 4) addressed a momentous question: Will climate change cause a large part of one of earth's two remaining ice sheets to disintegrate, precipitating a sea level rise sufficient to destroy much of worldwide coastal civilization?

Second, many of these assessments were very large, engaging hundreds of scientists and costing many millions of dollars. Indeed, they comprise a significant fraction of all work in earth and environmental science over the past several decades. The first phase alone of NAPAP, for example, lasted 10 years, cost nearly $600 million, and made major contributions to the scientific understanding of the relationship between industrial pollution, the hydrologic cycle, and lakes, forests, and soils. The Intergovernmental Panel on Climate Change (IPCC), discussed here in the context of assessment of WAIS, can be seen as the culmination of much of the earlier work in environmental assessment. The human and financial resources devoted to the IPCC over the past 31 years run into hundreds of thousands of person-hours and tens of millions of dollars of direct expenses. While one can argue about the efficacy of the IPCC in helping the world to prevent dangerous anthropogenic interference in the climate system, there is no doubt that the IPCC represents an important mechanism by which the world has tried to come to grips with this difficult, thorny, and potentially existential question and is thus worthy of study in its own right. It has also deeply affected the development of atmospheric science and related fields. We argue in this book that assessments do not just summarize existing knowledge but also create new knowledge and frame research agendas beyond the assessment. Scientific assessments have become a significant locus of scientific knowledge production and therefore are important to study along with fieldwork, laboratory practices, and other more familiar topics in science studies.

Third, we focused on these assessments because the three sets of epistemic communities involved overlapped to a significant degree, with many individuals being part of two and in some cases all three of these communities in one capacity or another. Taken together, these factors helped us to avoid the problem of making inferences across widely differing professional norms, although the range of expertise within any one of these assessments is very broad.

We make no claim that these assessments are representative of all assessments made everywhere in the world. In particular we are aware that there is an American bias to our study, both in the cases selected and in the questions that we have the most interest in addressing.

Our work would not have been possible without the crucial support of the National Science Foundation's Program on Science, Technology, and Society (although the results and opinions presented here are solely those of the authors). We are also indebted to the High Meadows Foundation and our home institutions over the course of the project: Princeton University; New York University; the University of California, San Diego; Harvard University; the College of Saint Benedict and Saint John's University; Indiana

University Bloomington; and the University of Alberta. We owe a special debt to those assessment participants who gave freely of their time to be interviewed, sometimes on multiple occasions. The book was much improved as a result of incisive comments by two anonymous reviewers, to whom we are grateful. We would like to thank our editors at the University of Chicago Press, first Christie Henry and then Karen Darling, for their support of the project. Finally, we would especially like to thank Kenneth Belitz, Benjamin Franta, Leonie Haimson, Neil Harris, Sheila Hellermann, Chunmei Li, Gene E. Likens, Elizabeth Lloyd, James McCarthy, John O'Reilly, V. Ramanathan, Martin Rees, Geoffrey Supran, Bob Watson, and Chris Whipple.

Abbreviations

AAAS	American Association for the Advancement of Science
AAOE	Airborne Antarctic Ozone Experiment (NASA)
ABM	antiballistic missile
AFEAS	Alternative Fluorocarbons Environmental Acceptability Study
AR4	IPCC *Fourth Assessment Report* (2007)
AR5	IPCC *Fifth Assessment Report* (2014)
CAA	US Clean Air Act
CAAA	US Clean Air Act Amendments
CCOL	Coordinating Committee on the Ozone Layer (UNEP)
CDAC	US Carbon Dioxide Assessment Committee
CEQ	US Council on Environmental Quality
CFC	chlorofluorocarbon
CFM	chlorofluoromethane (a class of CFCs)
CH_4	methane
CIAP	Climatic Impact Assessment Program (US Department of Transportation)
CISC	Committee on Impacts of Stratospheric Change (NAS)
CLIMAP	Climate Long-range Investigation, Mapping, and Prediction
ClO	chlorine monoxide
$ClONO_2$	chlorine nitrate, also written $ClNO_3$
ClO_x	oxides of chlorine
CLP	chlorine loading potential
CLRTAP	United Nations Economic Commission for Europe Convention on Long-Range Transboundary Air Pollution

CODATA	Committee on Data for Science and Technology Task Group on Chemical Kinetics
DDT	dichlorodiphenyltrichloroethane
DOE	US Department of Energy
DOI	US Department of the Interior
DPC	US Domestic Policy Council
EC	European Community
EDF	Environmental Defense Fund
EESC	equivalent effective stratospheric chlorine
EPA	US Environmental Protection Agency
FAA	US Federal Aviation Administration
FAR	IPCC *First Assessment Report* (1990)
FBI	US Federal Bureau of Investigation
GAO	US General Accounting Office (now Government Accountability Office)
GCM	global climate model
GRACE	Gravity Recovery and Climate Experiment
H_2SO_4	sulfuric acid
HCl	hydrochloric acid
HNO_3	nitric acid
HO_x	oxides of hydrogen
IEA	International Energy Agency
IEEE	Institute of Electrical and Electronics Engineers
IGCC	integrated gasification combined cycle
IGY	International Geophysical Year (1957–1958)
IIASA	International Institute for Applied Systems Analysis
IPCC	Intergovernmental Panel on Climate Change
NADP	US National Atmospheric Deposition Program
NAPAP	US National Acid Precipitation Assessment Program
NAS	US National Academy of Sciences
NASA	US National Aeronautics and Space Administration
NCAR	US National Center for Atmospheric Research
NOAA	US National Oceanic and Atmospheric Administration
NO_x	oxides of nitrogen
NOZE	National Ozone Expedition, August–September 1986 (NASA)
NOZE-2	National Ozone Expedition 2, August–September 1987 (NASA)
NRC	US National Research Council
NSF	US National Science Foundation
NSWS	US National Surface Water Survey

O	elemental oxygen, or a free oxygen atom
O_2	molecular oxygen (its usual form)
O_3	ozone (three oxygen atoms)
ODS	ozone-depleting substance
ODP	ozone depletion potential
OECD	Organisation for Economic Co-operation and Development
ORB	Oversight Review Board (NAPAP)
OTA	US Office of Technology Assessment
OTP	International Ozone Trends Panel
PSAC	US President's Science Advisory Committee
RADM	Regional Acid Deposition Model
RMCC	Canadian Federal/Provincial Research and Monitoring Committee
SAR	IPCC *Second Assessment Report* (1996)
SCAR	Scientific Committee on Antarctic Research
SIO	Scripps Institution of Oceanography
SLE	sea level equivalent
SNSF	Acid Rain—Effects on Forests and Fish (Norwegian research program)
SO_2	sulfur dioxide
SOS/T	State of Science/Technology (reports)
SRES	*IPCC Special Report on Emissions Scenarios*
SST	supersonic transport (airplane)
TAR	IPCC *Third Assessment Report* (2001)
UCSD	University of California, San Diego
UN	United Nations
UNEP	United Nations Environment Programme
UNFCCC	United Nations Framework Convention on Climate Change
USDA	US Department of Agriculture
UV	ultraviolet (radiation)
WAIS	West Antarctic Ice Sheet
WMO	World Meteorological Organization

CHAPTER ONE

The Need for Expert Judgment

INTRODUCTION

From ancient shamans, oracles, and diviners to the physicians of the World Health Organization and the scientists of the Intergovernmental Panel on Climate Change, there have long been individuals and groups with specialized knowledge who have been asked to provide judgment on issues that ordinary citizens and political and religious leaders felt unable to judge for themselves. In the twentieth century, the felt need for expert judgment grew, and institutionalized assessments of scientific knowledge for policy became a significant part of the landscape of scientific work and discourse. This book examines scientific assessments for public policy in the late twentieth and early twenty-first centuries. An obvious opening question is, Were there assessments before that time? If so, are recent assessments different? Exploring examples of expert judgment before the twentieth century, this chapter considers how assessments in the twentieth century can be distinguished from what went before.

THE PREHISTORY OF ASSESSMENTS

A vexing problem of the late medieval period was spirit discernment. Europeans in the Middle Ages faced the difficulty of distinguishing saints infused with the spirit of God from ordinary people possessed by demons. We might suppose that the two could scarcely be confused, but contemporaneous commentators agreed that the physical manifestations of divine and demonic possession were distressingly similar. Visions, trances, frenzies, levitation, the performance of miracles, feats of superhuman strength, xenoglossy, displays of stigmata, nudity, and other transgressions of social convention: these diverse manifestations were common to both. As historian Nancy Caciola has explained, the two kinds of spirit possession were "outwardly indistinguishable."[1] Moreover, this was no coincidence. As Paul had told us (2 Corinthians 11:14), Satan knew how to disguise himself as an angel of light.[2]

Divinely inspired prophets merited veneration, but false saints and demoniacs demanded condemnation, so it was essential to determine how to differentiate them. The challenge of discerning spirits was thus both epistemological and existential: epistemological because it involved questions of knowledge, existential because one's fate could rest upon it. The arduous work needed to differentiate between the two called for experts who examined cases, created criteria of discernment, and wrote reports, generating a large literature on the subject. In the fourteenth century Brigit of Sweden became a test case for spirit discernment when she was examined by a six-man panel: an archbishop, three bishops, a theologian, and an abbot. A panel of knowledgeable ecclesiastics also examined the case of Catherine of Siena, who died in 1380 of self-imposed starvation, and who, like Brigit, was subsequently canonized.[3]

Archbishops and abbots were not scientists—indeed, it would be many centuries before the term "scientist" would be coined—but they were individuals who had specialized knowledge relevant to the problem at hand. In that sense, they were experts whose views might inform action, including such weighty matters as canonization.

The rise of the modern nation-state brought new concerns and ways of constructing expertise. In eighteenth-century France, a group of prototechnocratic military engineers proposed a new form of artillery that could be produced faster and more cheaply through the use of interlocking and interchangeable gun parts. While perhaps less accurate and less durable than the heavy cannons they replaced, these new weapons could be quickly moved and mobilized, allowing French military strategists to think beyond the established rules of siege warfare and improve France's national security. This change meant rearranging the traditional relationships between the state, its armories, and the military. Armories now adopted a "systems" approach: traditional artisans were replaced by managers and planners who could arrange the work of machines and laborers into an organized whole. Social relationships were also restructured: titled lords who raised their own troops were now replaced by salaried professional officers who trained with designated artillery troops year round. Military leadership was now a career.[4]

A rival group of more traditionally minded military experts challenged these reforms. While no doubt motivated at least in part by a desire to protect traditional positions and privileges, they argued that the traditional techniques were more effective on the battlefield. To resolve this dispute, the minister of war convened a blue-ribbon panel of field marshals who had commanded French troops during the Seven Years' War (which raged from 1756 to 1763 and cost about one million lives). The panel—along with members of the public—witnessed a set of field exercises designed to demonstrate the efficacy of the new approach. The marshals sided with the reformers, and state policy was formulated to embrace the new approach.[5] Acknowledged expertise was now informing military policy.

THE RISE OF THE SCIENTIFIC AND TECHNOLOGICAL EXPERT

The nineteenth century witnessed a dramatic increase in activities that we now label scientific, as well as the increased visibility and social capital of savants who identified themselves by their disciplines—geologists, biologists, chemists, and the like. These "men of science" would soon come to be known collectively as scientists.[6] With the growth of science as a professional activity, these experts were increasingly called upon to resolve disputes that were understood to be both scientific and social and to produce reports of their findings. Here we may identify what we might consider to be early forms of the modern scientific assessment. Assessment in this context would mean any attempt to review the state of expert knowledge in relation to a specific question or problem, judge the quality of the available evidence, and offer findings relevant to the solution of the problem.[7]

In France, Louis Pasteur and Felix Pouchet argued about the nature of life and the fixity of species, with the latter advocating the theory of spontaneous generation and the former challenging it. Pasteur is popularly regarded as having debunked the theory of spontaneous generation via a strict adherence to scientific method, but both sides at the time offered experimental evidence in support of their claims. The Académie des Sciences deemed it so important to resolve this issue that it formed not one but two special commissions to judge the two sets of experiments.[8] While one Pasteur biographer bemoaned this approach as unsuitable to resolving a scientific dispute, the Académie did, in fact, settle the issue this way, siding with Pasteur and awarding him the Alhumbert Prize in 1862 for his experimental refutation of spontaneous generation theory.[9]

In the late nineteenth- and early twentieth-century United States, diphtheria was prevalent in urban areas, and recurrent outbreaks took the lives of many thousands of children. Historian Evelyn Hammonds notes that popular accounts generally suggest that the Pasteurian bacteriological model of disease led directly to new forms of medicine, including the use of antitoxins, but this is not in fact the case. When confronted with diphtheria epidemics in late nineteenth-century New York City, the medical community resisted both the bacteriological definition of disease and treatments whose justification rested on it. Partly, this was due to the medical profession's interest in maintaining its authority over disease as well as physicians' financial self-interest, but there was an epistemic issue at stake as well: at that time bacteriologists could neither account for nor control nonsymptomatic carriers of the diphtheria bacillus.[10]

Health department statistics showed a marked drop in mortality related to the use of diphtheria antitoxin, but many physicians remained skeptical. In 1896, the American Pediatric Society formed a committee to investigate antitoxin use.

The society's commission drew upon the clinical experience of 613 private physicians in its membership. After reviewing 3,384 cases, the commission ruled in favor of antitoxin and recommended that it be used in all cases as early as possible.[11] This still did not solve the problem of asymptomatic carriers but it did resolve the social question of whether physicians should embrace diphtheria antitoxin treatment, which at this point most did.

Vaccination was a major domain of expert assessment in the nineteenth century, because of both physician skepticism and public resistance. Historians and public health officials have noted that the British Vaccination Acts of the 1840s and 1850s—which mandated childhood vaccination for smallpox and outlawed variolation (the long-standing practice of exposing people to bodily fluids taken from a person with a live case of the disease)—were resisted by opponents who saw the acts as infringements on civil liberty. This resistance took the form of antivaccination leagues, protest, civil disobedience, and even riots. In response, a Royal Commission was established in 1885 to hear evidence for and against vaccines. The commission sat for seven years, during which it held 136 meetings, heard testimony from 187 witnesses, and examined two children suffering from ill health alleged to have been caused by smallpox vaccination; the final report extended to more than 500 pages.[12] Among those who testified against vaccination was Alfred Russel Wallace, the codiscoverer of the theory of evolution by natural selection, who argued that the recent observed decreased smallpox mortality was largely due to improvements in sanitation, not vaccination.[13]

The committee's charge was to consider both scientific and social questions regarding vaccination, though it did not sharply distinguish them in this way. The committee's members were scientific men, such as professor of anatomy and physiology and fellow of the Royal Society Michael Foster, but its chair was Farrer Herschell, a lawyer and lord chancellor of England. Scientifically, the committee asked whether there was a theoretical basis for believing that smallpox vaccination would be protective and, irrespective of theoretical understanding, whether there was sufficient empirical evidence to conclude that it is. Socially, the panel recognized the reality of objection and noncompliance. To discern and comprehend these social realities, the committee solicited extensive testimony from those whose objections were moral, philosophical, or personal. This led to a broadly framed discussion that included questions of compulsion and penalties for noncompliance. Among the topics discussed were the harsh treatment of parents by magistrates and the unfairness that ensued when parents who continued to refuse to vaccinate their children were repeatedly fined for what was, in effect, a single infraction. The commission's final report concluded that vaccines did protect against smallpox but recommended the abolition of penalties for noncompliance with the vaccination law. The new Vaccination Act of 1898 reflected this change and introduced a "conscience clause" allowing parents to decline vaccination on grounds of personal belief.[14]

As we will see in the chapters that follow, in the twentieth century, many scientists, legal scholars, and others would argue for a sharp separation between science and policy, but this distinction was not one about which the participants in the vaccination commission were unduly concerned.[15] Another difference between the vaccination report and most twentieth-century assessments is the inclusion of a detailed dissenting opinion. The 1885 Royal Commission Report includes a report of over 150 pages by the "dissentients," W. J. Collins and J. Allanson Picton.[16] The former was a physician, the latter an independent member of Parliament who had been accused of heresy for his unorthodox religious (and perhaps, as well, his radical political) views. Like Wallace, they were not persuaded that vaccination was the principal reason for declining smallpox mortality and therefore argued that it would be "unwise to attempt to enforce vaccination on those who regard it as useless and dangerous"—a position that the rest of the panel essentially accepted, insofar as they recommended abolishing penalties for noncompliance. However, the "dissentients" went further, arguing that "it would be simpler and more logical to abolish compulsory vaccination altogether."[17]

Social problems that required expertise to resolve often involved tensions over who had the relevant expertise and authority to lay claim to a particular domain. In Victorian-era Britain, Franz Mesmer's theory of animal magnetism, popularly known as mesmerism, was fashionable. Mesmerists claimed to be able to exert mental control over the minds and bodies of others and to use this power to cure psychological illnesses and anesthetize patients for surgery, sometimes simply by the laying on of hands. Popular audiences welcomed these claims, but they potentially undermined the authority of emerging scientific and medical professionals.[18] Moreover, the mid-nineteenth century was a time when a nonmedical perspective on madness—which saw it as a moral defect rather than a brain dysfunction—was threatening the medical profession.[19] To counter this, doctors were eager to find new theories and therapies that could be integrated within their own naturalistic paradigm of madness. Phrenology, for example, became one route through which doctors could explain moral defects in physical terms.

The acceptance and use of what many would today consider pseudoscience by physicians in the service of maintaining their expert authority makes it difficult to characterize the mesmerism debate as science versus pseudoscience. In fact, the matter of where to draw the line between science and pseudoscience, medicine and quackery, was settled not so much by knowledge but by disciplinary boundary work: the drawing of expert boundaries in a manner that relegated mesmerism to the fringe—defining away the mesmerist's expertise.[20] Here we see illuminated one feedback dynamic of knowledge production (discussed further in chapter 5): professional expertise helps to resolve social problems and these resolutions help to define what constitutes pertinent expertise. Socially acknowledged problems become the contested space within which professional groups define their collective identity, stake their professional claims, and forge agreement on what constitutes knowledge.[21]

As the category of "scientist" became solidified in the late nineteenth and early twentieth centuries, it also became codified as a recognized locus of specialized knowledge on which society could draw to help resolve contested questions. Questions regarding madness and normalcy, disease and health, and technology and its impacts came to be viewed increasingly as the domain of science, and so society increasingly turned to scientists to answer questions about them. The industrialized nation-state needed diverse forms of technical expertise in order to run its affairs; experts with knowledge were becoming increasingly viewed as important, even essential.[22] Scientists became the designated experts to help resolve a variety of societal problems, many of which were themselves consequences of science- and technology-inspired modernization. In the twentieth century, Lewis Mumford labeled these the questions of "technics and civilization."[23]

Technics were the focus of numerous commissions in the nineteenth and twentieth centuries in Europe and the United States. Typically, the impetus was failure: problems with steam engines, boats, and railroads, and especially collapsing bridges. Exploding boilers were a persistent and deadly problem in the steamboat industry, and in June of 1830 the newly founded Franklin Institute in Philadelphia empowered a committee of its members to investigate the causes of high-pressure boiler explosions. The committee eventually received funding from the US secretary of the treasury to support a set of experiments (the first grant of its kind) to understand the problem. For six years, University of Pennsylvania professor Alexander Dallas Bache directed a committee that blew up boilers in a quarry on the outskirts of Philadelphia. Based on the results of these experimental explosions, the Franklin Institute committee presented two reports to Congress, recommending guidelines on materials, design, construction, and maintenance procedures.[24] The reports were mostly ignored until President Van Buren urged the passage of legislation. On July 7, 1838, Congress passed a weak attempt at regulation, including watered-down versions of the reports' suggested guidelines. In 1852, Congress passed stronger legislation that established boards of inspectors to investigate infractions and accidents. Under this law, the owners of steamboat companies bore legal responsibility for the safety of their vessels. Driven to a significant extent by the work of technical experts, the US Congress acknowledged that industrial life required regulation to protect people, even if this meant intruding on private enterprise.[25]

THE INSTITUTIONALIZATION OF ASSESSMENT

This brief summary is by no means a comprehensive history of expert advice, but it is sufficient to demonstrate that experts in possession of specialized knowledge have long offered advice to those in power, that this advice has been both solicited and volunteered, and that, by the late nineteenth century, something similar to the contemporary scientific assessment had begun to emerge.

This should not come as a surprise: the existence of state-sanctioned honorific scientific bodies such as the French Académie des Sciences, the British

Royal Society, and the US National Academy of Sciences was predicated on the notion that such bodies not only would enhance the prestige of science in their countries but would benefit the state whenever specialized knowledge was required.[26] What (if anything) is different about the studies we address in this volume?

The types of work done by these commissions and committees before the twentieth century overlapped with the types of work done later; indeed, one might suggest that these early examples helped to establish the relationships we have today between expertise and governance. Moreover, there is no doubt that the individuals involved in these analyses performed a similar role to contemporary experts: they were seen to be in possession of specialized knowledge and were called upon to give advice based on that knowledge. In these respects, we may say that assessment is an old phenomenon, and the appeal to expertise in one form or another when faced with thorny problems has been persistent.

But there are some significant differences. One is institutional continuity. The expert commissions in our early examples were ad hoc: when a particular question was answered, the group disbanded. No lasting institutional apparatuses were constructed; individuals were addressing questions, not creating organizations that became bureaucratically instantiated and took on lives of their own. Nor had "assessment" been given that moniker. In short, while the precedent existed for expert intervention in societal problems, until the twentieth century no permanent infrastructure had been developed through which this kind of intervention could be enacted.

Perhaps for this reason, these earlier activities lacked many of the structural components that we observe in assessments today. Concerns about the structure and rules of assessment—including balance of interests and bias, conflict of interest, audience, and consensus building—are almost entirely absent in these earlier instances. Nor were they subject to peer review. Participants reviewed evidence, listened to testimony, and in some cases examined patients or performed experiments of their own and then passed judgment as eminent individuals. Without further historical investigation it is impossible to say how these activities were viewed by contemporary onlookers, but it appears to be the case that the reliability of the assessment was assumed to derive from the distinction of the individuals involved. And the assessment was the review; it did not have to be reviewed again.

However, we can discern in the nineteenth-century examples a suggestion of the institutional apparatus that would become formalized in the twentieth century. The Royal Commission on Vaccination was ad hoc—its members disbanded when the work was done—but they did work together for seven years, issuing five reports prior to their final one, and the idea of the royal commission persisted. By the end of the nineteenth century, the royal commission was a recognized mechanism for addressing socially important problems that involved technical expertise.

As assessments became institutionalized, they also grew in size, sometimes dramatically. Before the twentieth century, expert studies were typically conducted by a few people—often as few as three, rarely more than a dozen. The assessments that we study in this book involve dozens to hundreds of participants (and more if reviewers are counted). By the late twentieth century, large-scale, organized, and formalized assessments of the state of scientific knowledge had become a feature of the scientific landscape, a recognizable and regularized form of scientific work.

In the United States, we see a dramatic increase in the range, depth, and complexity of assessment after 1945, one that parallels the well-documented transformation of American science in the second half of the twentieth century.[27] This transformation included an increasingly close relationship between scientists and the federal government; a dramatic increase in funding for science (funding that among other things facilitated the development of the expanded earth and environmental sciences highlighted in this volume); a resultant growth of the scientific community overall; and an increased alignment of the focus of scientific investigations with the goals of the national security state, particularly during the Cold War years.[28] This period also saw increasingly conscious efforts by scientists to organize and direct American scientific research and increase science education in American schools and universities, through organizations like the newly founded National Science Foundation as well as older institutions such as the National Academy of Sciences and its National Research Council. The professional societies and associations of the various disciplines that had been growing and gaining authority since the interwar years also played a part in these developments. In short, the rise of assessments—both in fact and in name—coincides with the phenomenon that scientists and scholars have labeled "Big Science."[29]

The growing authority of science in the United States (and, arguably, elsewhere) and its organizations intersected with the emergence of what historian Brian Balogh has dubbed the "proministrative state." According to Balogh, the US government emerged from the Second World War a much more organized and powerful political actor than it had previously been. Not only did the postwar state require networks of expertise to administer its wider reach, but it was also now able to shape those networks and, with the cooperation of the professions, create and support well-organized groups of experts. This relationship was symbiotic: "Ultimately, it was the resources of the state—both financial and managerial—that the professionals could not do without; it was the prestige and the problem-solving capability of the professionals that tempted the state."[30] Recent studies of science during the Cold War have borne out this symbiosis and outlined the important ways in which national security concerns and scientific research programs became closely aligned during this period.[31] One part of that alignment was a growing allocation of resources to scientific research; another was a growing expectation on the part of the state that scientists would be available to answer important questions about national security and other matters.[32]

As a result of this rapidly evolving relationship between science and the American state during and after World War II, the US National Academy of Sciences and its National Research Council began to formalize their assessment activities. As historian Hunter Dupree observed, the academy had attempted to advise the government on scientific and technical matters since its formation in 1863.[33] Aside from its mobilization in times of war, however, these efforts bore little fruit. Moreover, more than a few of the academy's members questioned the wisdom of collectively taking on an advisory role. However, the academy's increased activities in World War II and in the postwar period led to a complete reorganization of the National Research Council as an operating arm of the academy in its formal relationship to the state and its agencies. What was once questioned was now taken for granted: that one of the academy's main purposes was to assess scientific and technical matters for the government.[34] By the time the assessments addressed in this book took place, the council had put in place a structured process that drew upon the expertise of scientists, engineers, and experts throughout the United States. The academy now touts its "consensus report" process as the gold standard in expert advice.

THE CATEGORY OF CONSENSUS

With the institutionalization of assessment came an increased focus on consensus. Figure 1.1 tracks the use in English of the terms "assessment" and "consensus": both were scarcely used before the twentieth century, and then their use rises dramatically, and more or less in tandem, after World War II. Consensus, it would seem, emerged in the mid-twentieth century as a preferred method for speaking scientific truth to power.[35]

The cases presented in this book (chapters 2, 3, and 4) show that achieving consensus is generally viewed as an important goal of the expert assessment

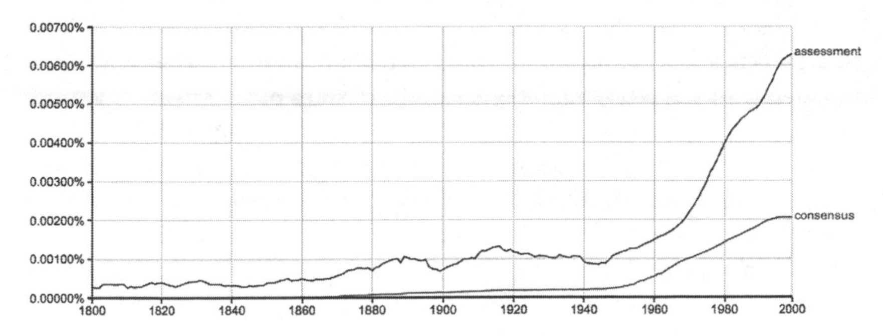

FIGURE 1.1 Google ngram tracing the relative frequency of "consensus" and "assessment" from 1800 to 2000.

process. Consensus appears to provide a way of signaling the agreement of experts about what knowledge is important enough and sufficiently settled to inform policy making—what Charles Kennel has called "decision-ready" knowledge.[36] It also allows scientists—or at least discrete groups of scientists—to speak with a collective voice. The unstated suggestion is that if this voice is univocal, then the science must be reasonably secure.[37] Conversely, if scientists cannot reach consensus, this might be viewed as a signal that the science is insufficiently settled to inform public policy.

How did consensus come to be a central component of assessment work? There are many models of deliberation: scientists could, for example, follow the model of the US Supreme Court, voting on decisions and offering diverse opinions supporting or dissenting from the majority view. As we have seen, the 1885 Royal Commission on Vaccination presented the dissenting as well as the majority opinion.

The appeal to consensus may be seen, at least in part, as the product of a situation in which governmentally supported science was dramatically expanding, not just in the United States but across the globe—yet the relationship between scientists and their patrons was neither clear nor uncontested.[38] In a world in which key relationships are in flux and the commitments of patrons are potentially insecure, it may be helpful not to highlight differences and dissent but instead to emphasize points of agreement and accommodation. It may also be that a consensus approach to the growing fields of expert knowledge (growing in terms of sheer volume as well as complexity) was what was most useful to the bureaucratic decision makers who came to rely upon expert advice.[39] Our findings suggest that scientists certainly think this is the case.

By the time scientists began to assess the scientific evidence regarding acid precipitation, stratospheric ozone depletion, and the risk of rapid disintegration of the West Antarctic Ice Sheet, the idea of consensus as a product of an assessment was already in place. Nonetheless, consensus did not become a "plug and play" technology; the cases presented in the following chapters illustrate significant differences in the ways in which consensus has been conceptualized. Institutional context determines many of the elements of assessments. The question of how consensus fits into the assessment process and what it is taken to mean is no exception.[40]

THE ROLE OF CONSENSUS IN OUR THREE CASE STUDIES

In our studies, the issue of consensus appeared most explicitly in ozone assessment, where we see scientists speaking self-consciously about consensus, both how it should be achieved and how to represent it to policy makers. Scientists involved in ozone assessment argued for the power of univocality, striving not just for consensus in any one particular report but also to create a single international process that would replace the diverse reports that had already been produced by various national agencies and organizations. They argued that the diversity of perspectives offered in this multiplicity of reports was problematic because it permitted politicians to make mischief: by playing up the disagreements between the reports and the limits to scientific knowledge made evident by their comparison, policy makers found an effective tactic for postponing action (chapter 3).

While areas of agreement across multiple reports could, in principle, have been taken to indicate a larger consensus than that offered by a single unequivocal report—and while scientists did push policy makers to focus on the areas of agreement among the several reports on ozone depletion that appeared between 1975 and 1985—the differences between the reports were used by some policy makers and industry representatives as evidence of a lack of consensus and therefore a warrant for delay. It was this observation that led Robert Watson, the creator and cochair of the international assessments that influenced the Montreal Protocol's assessment process, to adopt univocality as a central structuring principle. Consensus, along with internationality, emerged for Watson and his colleagues as an explicit goal in order to try to prevent disagreement from being used as an excuse for delay. The impulse toward univocality also arose in the debates on how to characterize the risk of rapid disintegration of the West Antarctic Ice Sheet. Scientists involved in that effort believed that it was important for them to come to some kind of agreement, despite what were, in fact, substantial differences of opinion about the character and imminence of that risk (chapter 4).

Watson wanted ozone assessments to be large, international, and highly inclusive, in order to underscore their consensual nature—what Edward Parson has called their "authoritative monopoly."[41] This (Watson hoped) would prevent skeptics from calling on other experts to support divergent claims. However, this inclusivity had its limits; among other things, Watson excluded or marginalized scientists who had made strong public statements about the need for ozone policy. The conspicuous example of this was F. Sherwood Rowland, who later won the Nobel Prize for his ozone work. Rowland was excluded from a leadership role in early international assessments because Watson worried that his public position on the need for chlorofluorocarbon (CFC) regulation might be seen by some stakeholders as compromising his objectivity—and therefore the perceived objectivity of any assessment in which Rowland participated. Watson's assessments were structured around not just the production of consensus but the production of a particular type of consensus: one that would be perceived as free from bias associated with having a public platform.

The exclusion of a world-renowned expert because he had already spoken publicly on the issue at hand raises several problems. First, and most obviously, it could exclude some of those most familiar with the problem and its potential implications—as it did in Rowland's case. Second, it might discourage scientists from speaking on issues with political implications even while *not* serving on an assessment—lest they later be disqualified—thus preventing scientists from pointing out potentially serious emergent problems and denying society information that it should have. Third, experience shows that simply not including an expert like Rowland does not assuage potential critics. Those determined to find a reason to object to science will do just that, as they did in the ozone example.[42]

Moreover, it is not clear how either the authority of a report or the unbiased character of its committee membership would dissuade a policy maker from disagreeing or stalling. Political opposition does not recede simply because the composition of an assessment is beyond reproach, real or imagined. Historically, there is little evidence to suggest that political actors who have opposed environmental regulation have suffered during their own careers for ignoring expert consensus. Indeed, some have capitalized on it.

The effort to eliminate bias by excluding individuals who have taken a public position may lead to a loss of important expertise, as scientists who are very close to the issue, and understand it well, are the ones who are most likely to feel compelled to speak publicly about it. This loss of expertise may also bias the report and reinforce scientific conservatism in the assessment output—a tendency toward "erring on the side of least drama,"[43] as scientists who have felt motivated or compelled to speak up are going to be among those who are most concerned. While some scientists may refrain from speaking in public because they are shy, feel uncomfortable in the limelight, or worry what colleagues will think if they take on a public role, it is also likely that scientists who judge the issue to be less worrisome will be less motivated to speak up than those who judge it to be more worrisome. Therefore, ceteris paribus, those who do speak up are likely to be more alarmed than those who do not, and their exclusion will skew the assessment toward underestimation of the severity of the problem. Thus the acknowledged need to avoid real or perceived political bias may lead to unacknowledged epistemic bias.

ASSESSMENTS AND ONGOING RESEARCH

Our study of the West Antarctic Ice Sheet (WAIS) assessments suggests that a robust research program is necessary for a successful assessment and meaningful consensus (whether the research takes place outside of the assessment, as is typically done, or within the same program, as in the US National Acid Precipitation Assessment Program, NAPAP); it also demonstrates that this research alone is not sufficient. There must also be a perceived threat or an information need that drives the research, assessment, and ultimate consensus in policy-relevant directions. This perception of threat is shared, at least to some degree, by both scientists and policy makers, and this creates a shared social space within which areas of uncertainty can be tied to information needs, and within which scientists can begin the deliberations from which a consensus will be articulated.

In the early years of WAIS assessments, the work of tying scientific uncertainties and research needs to policy-relevant questions was done by key individuals who regularly moved across a porous boundary between science and policy, and who had some idea what types of information were useful to policy makers. Roger Revelle served in this capacity for WAIS. He had already been inducted into the science/policy world through his activities as director of the Scripps Institution of Oceanography and his involvement in the International Geophysical Year, as science advisor to the secretary of the interior, and through his participation on the Governing Board of the National

Academy of Sciences/National Research Council. It was Revelle who pushed WAIS researchers to prioritize their research needs and tie them directly to the timescale of potential WAIS disintegration—what he saw as the policy-relevant dimension of WAIS disintegration.

Our WAIS study suggests that the felt need to come to consensus about the policy-relevant dimension of a scientific question helps to elucidate areas of policy-relevant uncertainty and research needs, and this shapes ongoing and future research. It does this not only by directing scientists toward these areas of uncertainty and motivating them to reduce the uncertainties but also by helping to align their existing research interests with the interests of policy makers.[44] In this way, the assessment process influences the direction and focus of research.

In the earliest WAIS assessments researchers came to the consensus that WAIS disintegration was not a threat in a politically relevant time frame, but there was enough disagreement and uncertainty to keep the issue alive as a policy-relevant area of research. Thus stressing uncertainty can be a means to keep alive a line of research that might otherwise be dropped. In some contexts, a consensus process may lead to a broadening of the research space in order to make room for reasonable outlier positions and may be a means to placate those who feel that their views—or at least some nuances of their views—have not been adequately addressed.

The assessments we studied all gravitated toward the identification and closure of "research gaps." These may be understood as products of efforts to come to consensus about what still needs to be known or what knowledge could potentially change the current understanding. In the ozone depletion assessment, continued research led to a new understanding of mechanisms that turned out to be crucial in explaining the appearance of the alarming Antarctic ozone hole. In the WAIS assessments, the workshops and group efforts to prioritize research needs and improve models led to new research and productive arguments that ultimately improved understanding of ice dynamics. Argument and debate from one workshop to the next also helped to bridge disciplinary compartmentalization. The workshops created a forum within which subsets of researchers with overlapping areas of expertise, or intellectual territory, were able to learn from one another and to address perceived deficiencies in their approaches.

In this sense, assessments may be viewed as examples of what historian of science Peter Galison has described as "trading zones"—physical or intellectual domains where researchers, despite their disciplinary differences or disagreements on method, can develop a common language and agree upon common concerns.[45] They differ from Galison's trading zones, however, in at least one significant way: the common language developed in a "successful" assessment—one that impacts policy—has to be such that it can be meaningful outside the assessment as well.[46] At the very least the summaries of the consensus must be intelligible to the intended audience (policy makers) and credible to relevant experts who did not participate in its articulation.

WHEN ASSESSMENTS GO "WRONG"

The desire to achieve consensus seems to be implicated in a serious problem that arose in both ozone and WAIS assessment: that the focus on settled knowledge led to the omission of questions and findings that changed the consensus when understood and incorporated in later assessments. In ozone assessment, this was the science of heterogeneous reactions, which was omitted from early assessments because it was wrongly assumed to be insignificant (chapter 3). In WAIS assessment, some modeling techniques were excluded because they were new and unverified, but they could have led to a better appraisal of future sea level rise (chapter 4). In both of these cases, the result was underestimation of the threat: the assessment made the problem seem less severe than scientists later concluded it was.

These examples show that a strong focus on consensus as settled knowledge—one that either excludes important but unsettled or controversial science or obscures disagreement over what science should be considered—can be detrimental. If consensus reports include only that knowledge that can be agreed upon by all participants—what we might consider "least common denominator" knowledge—then at best the stated conclusions may be weak, ambiguous, or watered down, and at worst they may be severely misleading.

On the other hand, a failure to articulate a significant body of agreed-upon knowledge can undermine the value of an assessment. The case in point is NAPAP, considered by many to be the least successful of the assessments studied here.[47] This was a massive interagency effort primarily focused on improving the state of knowledge about acid rain, its causes, and its effects on the environment, and providing a consensus report to Congress (chapter 2). NAPAP leaders seem not to have taken seriously the task of coming to a policy-relevant consensus—or, for that matter, even to have taken much interest in integrating the findings of the research programs supported under NAPAP. One reason this program has been deemed a failure is that it did not provide an official assessment report that influenced policy.

The institutional context of NAPAP may account for this failure to value and achieve consensus. The scientists and agency administrators involved in NAPAP were not motivated to articulate the existing consensus on acid rain. This may seem surprising, since the general consensus of most experts did not change significantly over the course of NAPAP. Scientists working outside NAPAP during this time believed the evidence was sufficient for policy action and emphasized this consensus, and much of the science that led to this conclusion came either directly or indirectly from NAPAP-funded research projects and initiatives. Yet NAPAP itself failed to make that case; NAPAP reports tended to highlight the unresolved issues and remaining uncertainties. This suggests that a research-oriented assessment can become focused on highlighting uncertainties at the expense of providing a policy-useful product.

CONCLUSION

The rise of consensus as both a category and a goal of assessments is closely tied to the history of assessments themselves. Consensus reports emerged in the mid-twentieth century as a means by which scientific experts sought to give clear and actionable advice to their own governments and the world at large. Consensus, as an ideal, has played a key role in notions of efficacy, as scientists acted on their belief in the power of univocality. Other approaches—such as expressing majority and minority views—were available and in some cases considered, but they did not prevail.

In taking the consensus approach, scientists in the mid-twentieth century perhaps intuitively perceived what social science research has since demonstrated: that expert disagreement, or even the appearance of it, can undermine public confidence in those experts and the science they are trying to communicate.[48] Smithson, for example, has demonstrated that conflicting estimates from experts generate more doubt and distrust in the minds of observers than agreed-upon estimates do, even if the range of the latter is as great as that of the former.[49] Similarly, Cabantous found that insurers assigned higher premiums to risks that were expressed by conflicting estimates than to risks that were expressed as consensual but equally uncertain.[50]

Whether one's goal is to generate trust in experts offering advice or confidence in the specific advice offered, it appears that univocality is likely to support that goal more effectively than polyvocality, however carefully the latter may be expressed. That said, the concept of consensus as displayed in our studies is not easily reducible to simple univocality: consensus appears to have taken multiple forms in various institutional contexts. Despite the emphasis assessors and their critics place on consensus, there is no singular established definition of consensus at work within assessments nor a universally accepted set of rules by which it should emerge. While the consensus report is often seen as the product through which a direct relationship between experts and policy makers is maintained, assessments may influence decision making in diverse ways. They introduce scientists to policy concerns, they help to guide research in policy-relevant directions, and they provide scientists and policy makers with areas of overlapping concern. Consensus is one element in this interaction, but not necessarily the most important.

Today, the US National Research Council produces over 200 assessment reports each year,[51] and similar reports are produced around the globe. Since the mid-twentieth century, various national and royal academies and international bodies have produced thousands of assessment reports on myriad subjects. As in the nineteenth century, many of these investigations have focused on technology—particularly technological failure—but others focus on environmental impacts, on the uses of science and technology in both the private and public sectors, on education, on public understanding of science, on science and the law, and other topics.

In the second half of the twentieth century, assessing science for societal decision making became a major form of scientific work. The modern scientific assessment is similar in many ways to earlier attempts by experts to sort out troubling issues but is distinguished by its scale—typically much larger—and its institutionalization. Yet at its core, assessment remains a kind of discernment, as experts gather and judge evidence and attempt to discriminate among diverse, competing, and sometimes conflicting claims. While we may no longer worry about the specific problem of discerning spirits, the general problem of discernment is very much with us, and equally existential.

CHAPTER TWO

Assessing Acid Rain in the United States: The National Acid Precipitation Assessment Program

INTRODUCTION

NAPAP was created by Congress in 1980 as an interagency program that involved 20 federal entities. In 1991, after 10 years of research and more than $550 million in funding, the US National Acid Precipitation Assessment Program (NAPAP) published its Integrated Assessment. NAPAP had the broadest scope of any analysis of an environmental issue for its time,[1] was probably the longest study and spent more money on research than any other acid rain program. (By comparison, the eight-year-long Norwegian acid rain research program, Acid Rain—Effects on Forests and Fish (SNSF), spent a total of $16 million.) Yet, despite its large size and its wealth of resources, NAPAP's assessment is widely considered to have been a failure because its published reports failed to influence policy makers.[2] While NAPAP formally continues to exist, it had its greatest influence in its first decade, and this is the period with which we are concerned.

A broadly held view is that the program focused on advancing knowledge about acid rain at the expense of producing a state-of-knowledge assessment relevant to policy making. Furthermore, NAPAP's critics point out that it published its results too late—one year after the passage of the 1990 Clean Air Act Amendments (CAAA). The literature also points to evidence of management failures and politicization that surfaced even before the program produced its final Integrated Assessment.[3] When NAPAP published an Interim Assessment in 1987 (already two years behind schedule) that attracted much criticism, the Canadian minister of the environment denounced it as "voodoo science."[4] NAPAP, its critics argued, failed to protect the scientific integrity of the assessment process.

This chapter argues that the story of NAPAP is more complex than the standard view suggests. To better understand NAPAP, we give close attention to the historical peculiarities of the program's development, organization, and structure. Between 1980 and 1990, NAPAP had three different directors: Chris Bernabo, a young and aspiring earth scientist; J. Laurence Kulp, a geochemist and senior scientist at Weyerhaeuser; and James R. Mahoney, a meteorologist and director at Bechtel Group, Inc. The management styles of the three directors differed considerably, as did their conceptions of the assessment process and its objectives, as well as their views on the necessity of acid rain legislation.[5] NAPAP was further influenced by the agendas and expectations of the federal agencies involved, in particular by the Environmental Protection Agency (EPA), the National Oceanic and Atmospheric Administration (NOAA), the Council on Environmental Quality (CEQ), the Department of Agriculture (USDA), the Department of Energy (DOE), and the Department of the Interior (DOI).

In addition to studying what went on within NAPAP, we also closely examine the changing political context in which NAPAP operated. NAPAP started under the Carter administration, was carried out primarily under the Reagan administration, and published its final assessment while George H. W. Bush was president. The expectations of decision makers about NAPAP were neither coherent nor consistent over the course of the decade between NAPAP's founding and the publication of its Integrated Assessment.

This chapter explores NAPAP's peculiar dynamics based on interviews, archival documents, and published NAPAP reports. It focuses primarily on three questions: First, how did scientists involved in NAPAP come to consensus, given that their assessment developed simultaneously with an overtly political debate? Second, how did the relationship between research and assessment within NAPAP develop over time? And third, is this relationship better understood as one of conflict or mutual support?

We begin with a brief historical overview of how acid rain emerged as an issue for scientific assessments. Then we discuss why and within what context NAPAP was created. We show that from the outset NAPAP was envisioned primarily as a research program, and we explain what factors led to the neglect of the assessment dimension (in the sense of informing policy). Finally, we describe the tensions between NAPAP's research and assessment dimensions and analyze the ways in which these tensions affected the production of the Interim and Integrated Assessments.

We find that, for much of NAPAP's existence, there was very little motivation within NAPAP to produce a final assessment (what we herein refer to as "push") and likewise no great desire on the part of most policy makers for an assessment (a corresponding "pull"). The fact that both push and pull were lacking in the case of NAPAP differentiates this assessment from the two others discussed in this volume. When the final report came due, producing a meaningful assessment proved difficult because of the amount and variety of compartmentalized research, a policy framework that was developed at the last minute, and an inadequate time frame. We also find, however, that NAPAP's research program, and even some of its assessment activities (in particular some of the modeling tools that it developed), did have an indirect and informal impact on the development of the 1990 CAAA and on future environmental and natural resource assessments. In the conclusion, we draw from the peculiar story of NAPAP some general observations about the factors that shape environmental assessments and their success or failure in influencing policy.

HOW ACID RAIN BECAME A TOPIC FOR SCIENTIFIC ASSESSMENTS

Acid Rain as an Environmental Problem

Acid rain emerged as a major environmental problem in the late 1960s. Although acid rain can occur naturally, for instance resulting from the chemical conversion of sulfur dioxide (SO_2) emissions from volcanoes into sulfuric acid, the term refers predominantly to man-made sulfur dioxide and nitrogen oxide (NO_x) emissions that react in the atmosphere to form sulfuric or nitric acid. So defined, the observation of acid rain can be traced back to 1852, when the Scottish chemist Robert Angus Smith (1817–1884) related damaging effects on materials and health to free sulfuric acid in the air in and near the industrial city of Manchester.[6] Smith noted "illness elevation, or depression of some kind" in relation to breathing air containing chemicals such as sulfuric acid,[7] and pointed out several effects on materials including the fading of colors, corrosion of metals, and crumbling of stone.[8] Other effects of acid rain that were noticed later in the twentieth century include declining fish populations and damage to and death of trees.

Yet acid rain was never a clearly defined topic. The natural acidity of rain can vary widely from place to place and over time, and the choice of a reference value for determining rain's acidity depends on the understanding of these variations in rainwater pH. Narrower definitions of acid rain, such as rain or snow with a low pH value, coexisted with broader definitions of acid deposition, including dry deposition in particulate form, which made up a considerable amount of the total acid deposited onto lakes, forests, and buildings.

Not only the definition but the conceptualization of acid rain as an environmental problem differed over time. In the late nineteenth and early twentieth centuries, acid rain was framed as a problem of local air pollution. In the 1970s, acid rain was primarily described as a regional problem, occurring in several parts of Europe, Japan, the northeastern United States, and southeastern Canada.[9] Recently, the US National Atmospheric Deposition Program (NADP, a precipitation monitoring program) defined acid rain as a global problem,[10] noting that pollutants from China can potentially travel to the United States.

Acid rain emerged as a regional and international problem in the 1950s due to the identification of long-range transport of air pollution as an important phenomenon. Long-range air pollution transport became prevalent due to increased burning of fossil fuels and the invention of tall smokestacks.[11] While in 1950 more than 75% of SO_2 emissions in the United States came from smokestacks shorter than 100 meters, by 1980 the number of short stacks had been reduced to 5%, and almost 60% of emissions came from stacks over 200 meters high.[12] Building taller stacks was a policy measure intended to help reduce pollution close to emission sources, but as emissions occurred at higher elevations and were dispersed over a larger area than before, pollutants remained in the atmosphere longer, had more time to react with other chemical species, and affected areas farther from the source.[13] Instead of solving the air pollution problem by distributing it over broad areas where (it was hoped) it would become diluted, acid deposition increased and affected more areas.

Acid deposition thus became a regional and international environmental problem, requiring not only new measures of abatement and control but also new and more complex scientific approaches to address its consequences. The changing chemistry of precipitation and its effects on terrestrial and aquatic ecosystems became a topic for systematic research in Scandinavia around 1955 and in North America around 1966.[14] Several disciplines and scientific fields such as limnology, agriculture, and atmospheric chemistry dealt with aspects of acid deposition, and approaches to acid deposition varied both within and between scientific disciplines. Disciplinary approaches to acid rain can roughly be divided into two areas: limnologists, forest ecologists, and crop scientists focused on the damaging effects of acid deposition in distinct ecosystems, while atmospheric scientists focused on the precursors of acid deposition and the formation of acids in the atmosphere. These disciplinary differences led to significant tensions in NAPAP's interdisciplinary assessment.

Systematic monitoring of European air chemistry started in 1947, when Sweden set up a number of precipitation measurement stations. During the 1950s, the program was extended to 11 other European countries and expanded to include more pollutants. In 1968, a first analysis of data from the European Air Chemistry Network demonstrated the annual expansion of the extent of areas receiving acid deposition in Europe.[15]

The Swedish scientist Svante Oden (1924–1986) played a central role in uniting the different scientific branches that deal with aspects of acid deposition. His popular articles and talks brought acid rain to the public's attention, first in Sweden, then in Europe, Canada, and the United States. In a 1967 newspaper article, "Nederbördens Försurning" (The Acidification of Precipitation), he described acid rain as "chemical war" between countries.[16]

The regional character of acid rain influenced how countries responded to the problem. The governments of countries that suspected they were receiving substantial amounts of acid deposition from other countries were among the first to fund research, conduct assessments, and implement control legislation. A Swedish proposal adopted in 1968 by the United Nations (UN) General Assembly led to the 1972 UN Conference on the Human Environment in Stockholm, where the topic of acid rain found its first international forum.[17] Norway's interdisciplinary SNSF research program, which ran from 1972 to 1980, was the first large-scale national study on acid rain. In 1972, the Organisation for Economic Co-operation and Development (OECD) started to focus on long-range transboundary air pollution and conducted a five-year study titled "Co-operative Technical Programme to Measure the Long Range Transport of Air Pollutants"[18] with "special attention being paid to the question of acidity in precipitation."[19]

In 1979, the UN Economic Commission for Europe Convention on Long-Range Transboundary Air Pollution (CLRTAP) was signed by 34 countries (including Canada and the United States as well as many European countries and the European Community). States recognized "the existence of possible adverse effects, in the short and long term, of air pollution including transboundary air pollution" and agreed to "endeavor to limit and, as far as possible, gradually reduce and prevent air pollution."[20] Protocols to the convention, in particular the 1985 Helsinki Protocol on the Reduction of Sulphur Emissions or Their Transboundary Fluxes, which came into force in 1987, established binding emissions reductions of at least 30% from a

1980 baseline. However, while 17 European countries and Canada signed the Helsinki Protocol, the United States, the United Kingdom, Poland, and Spain, all of which were major emitters, did not.

Early North American and European Acid Rain Assessments

At the time of NAPAP's creation in 1980, the Norwegian government had just published the results of its eight-year study of acid precipitation. Three years earlier, the OECD had released its first comprehensive report on long-range transboundary air pollution—the culmination of five years of research. Why did the US acid rain program start years later than its European counterparts? It was not that US scientists did not know about acid rain in the 1960s and 1970s. Precipitation measurements at the Hubbard Brook Experimental Forest, established in 1955 as a major center of hydrological research in New England, showed very low pH values (i.e., extreme acidity) in the 1960s. Papers by Gene Likens, Hermann F. Bormann, and Noye Johnson[21] and others pointed out that acid rain posed an environmental problem in the United States. The US National Academy of Sciences and the USDA Forest Service published several reports that dealt with aspects of acid deposition in the late 1970s,[22] and the Multistate Atmospheric Power Production Pollution Study published its results in 1979.[23] The United States lagged in initiating a coordinated acid rain research and assessment program because political information needs emerged later in the United States than in highly affected countries such as Norway or Sweden.

In the case of acid rain assessments, views differed widely on their expected contributions to the policy-making process. The SNSF program was an "interdisciplinary research program," and its two objectives were to "establish as precisely as possible the effects of acid precipitation on forest and freshwater fish" and to "investigate the effects of air pollutants on soil, vegetation and water, required to satisfy point 1."[24] It did not include policy options or policy recommendations. NAPAP had the broader task "to identify the causes and effects of acid precipitation" and "to identify actions to limit or ameliorate the harmful effects of acid precipitation."[25] The first objective established a comprehensive research program, and the second called for an assessment of the state of knowledge in order to provide policy makers with alternative policy options (see figure 2.1).

While NAPAP developed policy options but not policy recommendations, the working groups established by the 1980 Memorandum of Intent Concerning Transboundary Air Pollution between the United States and Canada

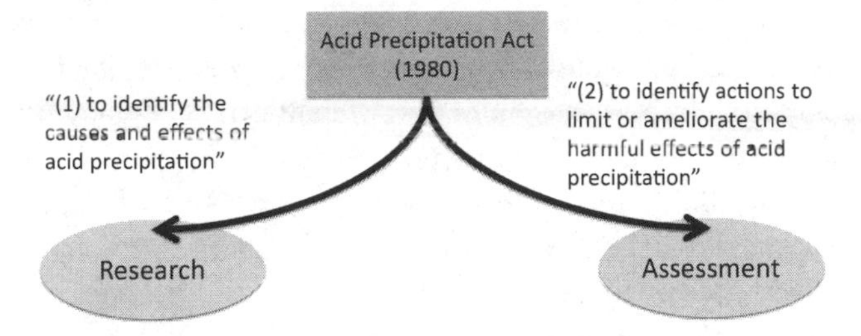

FIGURE 2.1 The two fundamental goals of NAPAP.

(which also assessed the problem) had, among other objectives, the task to provide policy recommendations by "propos[ing] reductions in the air pollutant deposition rates . . . which would be necessary to protect identified sensitive areas."[26]

In contrast to the assessments of ozone and climate change discussed in other chapters, there was no broadly inclusive international acid rain assessment; rather, assessments were conducted nationally, binationally, or regionally. However, areas of consensus and uncertainty about the causes and effects of acid rain existed in North America and Europe in the years before NAPAP initiated its research program. If we compare seven acid rain assessments (three from the United States, three from Europe, and one from Canada) that were published between 1976 and 1981 (see appendix 2.1 for a list of issues compared), we see that all assessments agreed that acidic deposition was largely man-made and that long-range transport of air pollutants that generate acid rain (precursors) and the acids themselves occurred.[27]

There was also a broad consensus that there was evidence of damage to aquatic ecosystems and of a historical trend in the emissions of SO_2. A controversial topic was whether source-receptor relationships could be established—that is, whether areas receiving acid deposition could be related to specific areas of emissions. The three European assessments[28] stated that such relationships could be established, while the North American assessments either did not specifically address this issue[29] or concluded that they did not have enough evidence to establish source-receptor relationships.[30] This can be explained to some extent by the fact that the National Research Council of Canada and the US National Research Council (NRC) assessments focused on effects on ecosystems rather than on atmospheric processes—an omission that foreshadowed a bigger controversy during the 1980s (the "linearity" issue, discussed below).

The most uncertain issue around 1980 was whether acid deposition caused substantial damage to forests. All five assessments that addressed this issue concluded that there was no or insufficient evidence of adverse effects on forests, but they also noted that adverse effects were possible. There was no difference in the evaluation of impacts on forests across European and North American assessments.

Only one multinational and one US assessment addressed the reliability of computer models of atmospheric transport of pollution, but this was controversial. The OECD (1977) regarded them as reliable, while the Interagency Task Force (1981) did not. This disagreement about the reliability of available models was closely related to the question of whether source-receptor relationships could be established. It became a more broadly discussed topic during the 1980s and emerged as one of the main differences between European and Canadian assessments compared to those conducted in the United States. Only two assessments[31] addressed the question of whether acid rain required government action, and both recommended immediate action.

The Differences between European and US Acid Rain Policy

At first glance it seems that acid rain regulation would have been more easily agreed upon in the North American setting, where negotiations involved only two countries, but the complex political dynamics of the European system in fact facilitated the earlier implementation of regulation in Europe. Due to the influence of Germany and the United Kingdom, the European Community (EC) was the main opponent to binding emissions reductions during the CLRTAP negotiations. Yet because of the active role of the Scandinavian countries, acid rain was on the EC's agenda. When Germany revised its policy in the early 1980s and supported acid rain regulation on an EC level, the political dynamics changed. Public concerns about forest decline in 1981–1982 led the Social Democrat/Free Democrat coalition to adopt emissions regulations for large combustion sources. Since Germany imported more than half of its acid deposition, it aimed at tackling acid deposition through the United Nations Economic Commission for Europe and the EC. However, the main piece of European acid rain regulation, the large combustion plants directive, did not pass until 1988 because the United Kingdom's Thatcher government had fiercely opposed emissions reductions in the previous years.

The US situation differed significantly. In contrast to Europe, acid rain policy in the United States, in addition to being a matter of domestic policy, became a bilateral issue at the international level, between the United

States and Canada. In his second Environmental Message (August 2, 1979), President Carter described acid rain as one of the most serious environmental problems facing the nation. Carter created the Acid Rain Coordinating Committee in 1979, in an effort to develop a coordinated federal acid rain program. Not only did the Carter administration take initiative on the national level, but the United States and Canada started to negotiate over transboundary air pollution in October 1978. The 1980 memorandum of intent between the United States and Canada had as its major objective the development of a bilateral agreement to combat transboundary air pollution.

In the United States, the Clean Air Act (CAA) of 1970 provided the regulatory framework for air pollution. The act included provisions for the use of technological control devices for new pollution sources, the protection of clean air regions, and the regulation of urban air quality.[32] However, it had not yet been used to address the long-range transport of air pollution or any adverse effects other than those on human health. The 1977 amendments to the CAA addressed the problem of interstate air pollution by including a provision that the EPA should not approve state implementation plans (originally aimed at reducing local health impacts of air pollution) that would "prevent attainment or maintenance" of air quality standards in another state.[33] Section 115 of the CAA dealt with international air pollution. States could be required to revise their air quality plans if their emissions "cause or contribute to air pollution which may reasonably be anticipated to endanger public health or welfare in a foreign country."[34] A precondition was that the foreign country must give the United States a similar right, which Canada did in December 1980.[35] In January 1981, EPA administrator Douglas Costle stated that there was sufficient evidence that US emissions were causing substantial damage in the Canadian environment and that "Section 115 authority could appropriately be used to develop solutions."[36]

Civil society organizations also supported curbing emissions. Major environmental groups such as the Environmental Defense Fund (EDF)[37] and the Natural Resources Defense Council pushed for acid rain regulation. During the 1970s, these groups advocated immediate action and filed several lawsuits seeking to curb emissions. While these developments seemed encouraging, two events then stifled US action on acid rain. The first setback was the oil supply crisis of 1979; the second was the election of Ronald Reagan.

In the context of the oil crisis, energy security moved high on the political agenda. In response, the administration planned to let power plants use more domestic coal (which generally had a higher sulfur content than the fuels it replaced) in place of oil or natural gas. Thus the emergence of acid rain on the political agenda became intertwined with energy security. Several congressmen raised the issue of increasing acid rain in the lengthy negotiations on the Energy Security Act. In this context, Senator Daniel Patrick Moynihan (1927–2003), a Democrat from New York, argued for a comprehensive interagency study,[38] and political support began to emerge for an organized assessment of acid rain. Then, in 1980, Ronald Reagan was elected president on a platform of reducing government regulation. In this context, some policy makers viewed funding of more research as a useful tactic to postpone action, a political strategy that dominated the Reagan administration's position on acid rain.[39]

The Political and Scientific Forces That Led to NAPAP

NAPAP emerged out of an alignment of interests of two influential groups: scientists and decision makers who asked for more research. From the perspective of scientists, more research would yield a better understanding of the extent of acid deposition in the United States (where nothing comparable to the European Air Chemistry Network yet existed) and its effects on aquatic ecosystems such as lakes and terrestrial ecosystems such as forests. More research also meant more funding for scientists and for agencies.

While all scientists agreed that more research would contribute to better understanding of the extent and effects of acid rain in the United States, they had differing views on whether the question of implementing regulation required more research. Gene Likens and Herbert Bormann argued in several articles in the early 1970s that even though data on the extent of acid deposition in the United States were scarce and "only some of the ecological and economic effects of this widespread introduction of strong acids into natural systems are known at present," policy makers should consider these effects "in proposals for new energy sources and in the development of air quality emission standards."[40] However, other scientists held that more knowledge was needed as a basis for regulation. A 1978 assessment prepared by four university scientists (James Galloway, Ellis B. Cowling, Eville Gorham and William McFee) at the request of the CEQ emphasized: "Although it is known that the acidity of precipitation has increased, the relative contributions of the acids causing this increase (sulfuric and nitric) are not known. Until these are known, effective control measures will be difficult to establish."[41]

The former view—that the state of available knowledge was sufficient to act—did not prevail in the political debate. In contrast to several European countries where policy makers acted on the basis of the state of knowledge that was available around 1980, the view that regulation required more knowledge dominated the US debate (at least among policy makers). The US case differed primarily due to the adversarial and legalistic approach to environmental regulation and the stronger influence of industry on the policymaking process. In contrast to the Netherlands, Sweden, and Norway, where small groups of scientists and policy makers collaborated closely on acid rain regulation,[42] US scientists knew that their findings would likely be scrutinized in the courts.[43] In addition, congressmen from states that were exposed to acid deposition, such as New York, where researchers reported on lake acidification in the Adirondacks, had an interest in determining more precisely the extent of acidification and the patterns of emissions in order to create and implement regulation. Moreover, Congress decided to create an interagency assessment because it wanted a consensus report from the otherwise independent agencies; as NAPAP's first director (1980–1985), Chris Bernabo, emphasized:

> The problem from a Congressional standpoint was if you had a hearing, you would have 12 different federal departments and agencies each having their own views and programs because there was no coordination. They were coming for funding for the same thing, with competing programs. Having a mandated interagency program forced them to integrate their efforts and come to consensus.[44]

NAPAP'S PECULIAR DYNAMICS

Establishing NAPAP

NAPAP was created to encompass research, monitoring, evaluation of impact, and policy recommendations, the latter largely based on NAPAP research but also including outside research from the United States and elsewhere.[45] The research and the policy dimensions were supposed to support each other, but this hybrid construction ultimately failed.

Congress established NAPAP under the Acid Precipitation Act of 1980 (Public Law 96-294), Title VII of the Energy Security Act, which President Carter signed into law on June 30, 1980. The act also authorized financial support from Congress and established an interagency task force that consisted of 20 members, jointly chaired by the secretary of agriculture, the administrator of the EPA, and the administrator of NOAA. Task Force members included representatives from the DOI, the Department of Health and Human Services, the Department of Commerce, the DOE, the Department of State, the National Aeronautics and Space Administration, the CEQ, the

National Science Foundation, the Tennessee Valley Authority, the directors of the Argonne, Brookhaven, Oak Ridge, and Pacific Northwest National Laboratories, and four presidential appointees.[46]

The task force defined NAPAP as a "program of policy-oriented research" with a goal "to identify the sources, causes and processes involved in acid precipitation," and to evaluate "the environmental, social, and economic effects of acid precipitation." The task force expected NAPAP to issue reports on the state of knowledge about acid deposition and its effects but also "recommendations about what policies and actions may be effective for managing acid deposition" and suggestions of "strategies for ameliorating harmful effects associated with acid precipitation."[47] As outlined in §8903 of the Acid Precipitation Act, NAPAP had nine main areas of activity:

(1) identifying the sources of atmospheric emissions contributing to acid precipitation;
(2) establishing and operating a nationwide long-term monitoring network to detect and measure levels of acid precipitation;
(3) conducting research in atmospheric physics and chemistry to facilitate understanding of the processes by which atmospheric emissions are transformed into acid precipitation;
(4) developing and applying atmospheric transport models to enable prediction of long-range transport of substances causing acid precipitation;
(5) defining geographic areas of impact through deposition monitoring and identifying sensitive areas and areas at risk;
(6) broadening impact databases by collecting existing data on water and soil chemistry and through temporal trend analysis;
(7) developing dose-response functions with respect to soils, soil organisms, aquatic and amphibious organisms, crop plants, and forest plants;
(8) establishing and carrying out system studies with respect to plant physiology, aquatic ecosystems, soil chemistry systems, soil microbial systems, and forest ecosystems;
(9) providing economic assessments of
  (a) the environmental impacts caused by acid precipitation on crops, forests, fisheries, and recreational and aesthetic resources and structures, and
  (b) alternative technologies to remedy or otherwise ameliorate the harmful effects that may result from acid precipitation.

In addition, the act charged the program with documenting and coordinating federal acid rain–related research and cooperating with other nations.

The task force would manage financial resources committed to federal acid precipitation research and development, as well as the technical aspects of these activities. This included the establishment of peer review procedures and periodic reports to Congress and the agencies.

NAPAP's Organization

The participating agencies crucially influenced NAPAP—and this influence ultimately hampered the assessment. The program's director had little independence from the agencies, the scientists primarily represented their agencies' missions and not the priorities of the program, and the agencies had no strong interest in conducting an assessment.

The highest organizational body of NAPAP, the Joint Chairs Council, included EPA, CEQ, DOI, USDA, NOAA, and DOE. The Joint Chairs appointed the director of the program and signed off on NAPAP's reports and assessments. At the base of NAPAP's organization were 10 (7 after 1985) task groups, each of which dealt with a specific aspect of acid rain and carried out "detailed planning and work" in its respective field.[48] One agency led each of the task groups, but they could include scientists from other agencies. They differed in their size and composition and had no formal membership. The respective task group leaders decided which scientists should be included.[49] The task groups' major responsibilities did not include conducting research but did include integrating the various findings of the different studies funded by NAPAP and compiling the scientific reports. The scope of the task groups, along with their composition, changed after NAPAP's reorganization in 1985 (see figures 2.2 and 2.3).

While most of the interviewed NAPAP participants had a positive view of the interagency cooperation in this program, some interviewees (e.g., Mahoney) reported tensions between some agencies, in particular between EPA and DOE:

> For the most part I would say that the differences that I would see would be that of the personal interests of the agency leaders and not so much [of the agencies]. . . . Well, EPA had the responsibility for the environment so you expect them to be, well, protecting the environment, and DOE was pretty strong of the other view, but DOE had a lot of very solid environmental thinking people too. . . . Sometimes the DOE people were very concerned about people they felt were too close to EPA, because they feared they'd be captured by the EPA point of view. . . . The sharpest differences were between DOE and EPA, generally speaking, not as much between Interior and Agriculture and the like.[50]

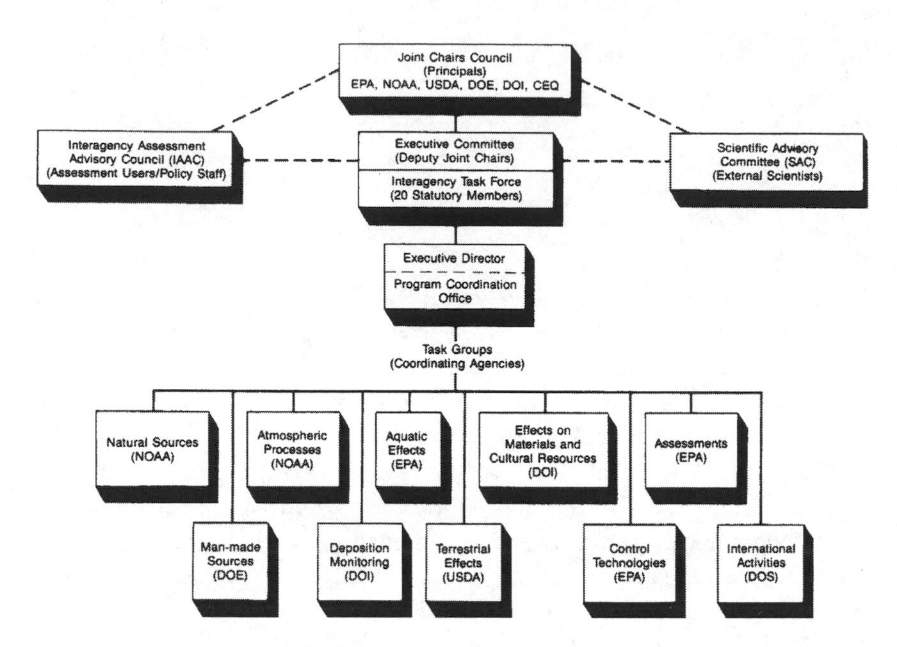

FIGURE 2.2 NAPAP's organizational structure before 1985 (reproduced from US General Accounting Office, *Acid Rain: Delays and Management Changes*, fig. 1.2).

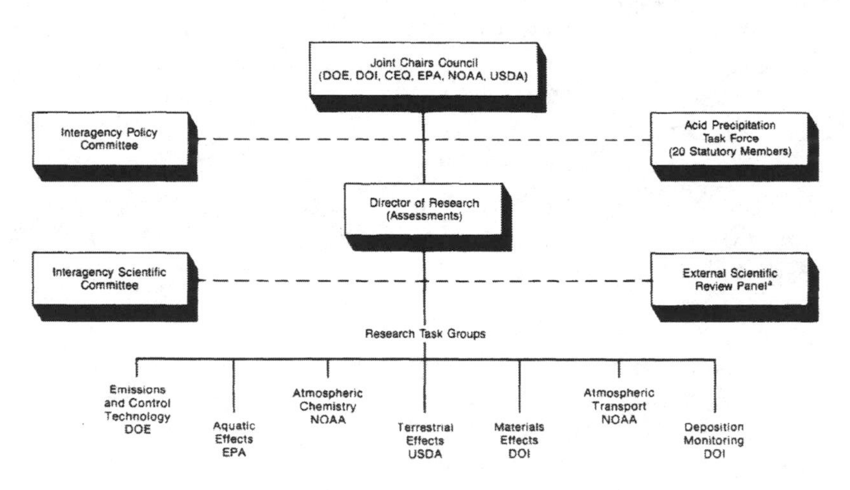

FIGURE 2.3 NAPAP's organizational structure after 1985 (reproduced from US General Accounting Office, *Acid Rain: Delays and Management Changes*, fig. 1.3).

John Malanchuk, a watershed modeler who joined the EPA in 1984 and who later became the deputy director for the Integrated Assessment, also reported tensions between DOE and EPA:

> Was DOE trying to be scientific and being careful in their analysis and their use of uncertainty or were they just trying to get at a level of resolution where they could attack everything and pull it apart? And the answer is some of each. DOE was definitely doing things like that and on the side of the white knight [EPA], there were people doing that as well. . . . Just because that's the way policy gets handed out, I guess.[51]

The different priorities of the agencies became very apparent when it came time to write the chapters for the research reports of the Interim Assessment.

The chapters of the Interim Assessment went through an agency review process, and the authors received differing feedback from the respective agencies. Malanchuk described the influence of the agency reviews on the assessment in the process of compiling the chapter on aquatic effects for the Interim Assessment:

> You would get into situations all the time that EPA makes a comment on a draft that I and Rob [Robert Turner, then at Oak Ridge National Laboratory] wrote, and we would make the changes because we didn't really disagree with them. Then the next draft would go out, and DOE would come back, and they would specifically make comments on exactly the changes that EPA had made on the previous draft. . . . This becomes untenable: you can't stand it, because the only thing that you can do is get people in a room and fight; you can't do it in a vacuum, because I don't necessarily disagree with what EPA wanted, and then when you talk to DOE, then DOE says, "Well, it isn't most lakes are acidic, because most applies to a majority. . . . It's not even many lakes, it is just some lakes." . . . And do I really care that much? The answer is no. So, OK, I make your change. But as soon as EPA sees it, they go nuts! It's trivial, and it's not necessarily wrong; it's more like: who cares the most.[52]

These tensions affected more than the report writing and review process. The different views of DOE and EPA not only challenged the authors of the assessment, they also had structural effects upon the program itself. One agency oversaw each of the major NAPAP-funded projects, such as the emissions inventory, the National Surface Water Survey (NSWS), or the Regional Acid Deposition Model (RADM). More often than not, the lead agency followed its own research interests. The agencies considered the compatibility of one agency's NAPAP research with that of another only as a secondary concern. This became a major problem for NAPAP when in 1990 the program had to produce its Integrated Assessment (see below).

NAPAP developed its first research plan in 1980 and 1981 but initiated research only in 1982.[53] Most major NAPAP-funded projects started in the years 1982 to 1985. However, in addition to the big research projects such as the NSWS, NAPAP funded several hundred smaller research proposals. The hundreds of scientists who received research funding through the agencies continued to work at their respective universities, agencies, or laboratories.[54] While it is beyond the scope of this chapter to explore the effects of this large-scale assessment on the research community, it is likely that NAPAP had a substantial impact on US acid rain research through its funding decisions.

Inside NAPAP: Research Instead of Assessment

From the beginning, most of NAPAP's participants envisioned it primarily as a research program. This was because the main groups of actors involved in the program—the agencies and agency scientists—had no strong interest in the policy dimensions. That NAPAP was, by design, a long-term program, where the policy evaluation would be based on research findings that had yet to be produced, also contributed to this bias toward research.

NAPAP's first director, however, had a different view on the importance of policy and—at least on paper—set NAPAP on a course with a clear policy orientation and an ambitious outlook overall. One of the 10 original task groups was asked to develop an assessment methodology and conduct the assessment. The program was supposed "to construct the means for comprehensive benefit-cost assessments," and the Assessment Task Group planned to work toward "a detailed accounting of economic damages."[55] The Assessment Task Group was also in charge of producing the Interim Assessment, scheduled for 1985. The 1983 annual report, however, emphasized that, to date, "the state of the science will not allow assertive recommendations. Trends are weak and evasive. Data are spotty." The assessment faced uncertainties related to fundamental questions such as the extent of damage caused by acid deposition and its rate of change.[56]

In 1983, the key questions for the assessment were

1. What is the physical, biological significance of current and expected adverse or beneficial effects from the deposition of acidic and acidifying materials in North America?

2. How are the composition and distribution of acid deposition in North America linked with emission patterns, and what significance do uncertainties have for control or mitigation strategies?
3. From what range of strategies for integrated emission control and receptor-oriented mitigation can policy makers choose?
4. Which strategies show the greatest promise of cost-effectiveness and what bounds of uncertainty should be placed around such conclusions?
5. What specific research would most effectively reduce the physical, biological, and economic uncertainties that decision makers must face in choosing among strategy options for dealing with acid deposition?[57]

However, in the early years, rather than evaluating the impacts of acid rain NAPAP was conducting research on assessment methodology. In the years 1982 to 1985 the EPA funded several research projects in the areas of "Integrated Assessment Methodology Development,"[58] "Coordination and Integration," and "Applications and Policy Analysis."[59] However, NAPAP never actually implemented these methods.

A bigger problem was that the EPA, within which the Assessment Task Group was based, had no strong interest in developing an Integrated Assessment framework. Instead, the agency was primarily interested in pursuing new research. Unlike ozone, in which scientists and agencies had a strong interest in doing an assessment, the agencies, agency scientists, and researchers whom NAPAP funded concentrated on the new research opportunities at the expense of the assessment: "Did people just take the money to do the science they loved to do? And write papers they loved to write? And worry about the assessment later? Because they had to in order to get the money? Sure, of course," said Malanchuk in an interview.[60]

One reason for these differences in the behavior of agencies and agency scientists in the case of the acid rain and ozone assessments could be that, in the view of those who conducted the assessments, acid rain did not pose as great a threat as ozone depletion. They may also have perceived acid rain as less of a global problem and more of a national concern. Moreover, the political climate during the Reagan administration supported NAPAP's research orientation.

NAPAP and US Acid Rain Politics

Although the program was approved under the Carter administration, NAPAP's work started after the Reagan administration came into office in January 1981.

This change in administration altered the political context of NAPAP significantly. The Reagan administration took the position that the scientific uncertainties were too large to implement emissions control legislation and that the available evidence suggested that acid rain was not a big environmental problem.[61] A Domestic Policy Council (DPC) memorandum for the president from January 1988 stated that "current US acid rain policy is adequate and that changes would be premature and potentially damaging to our economy."[62] In the 1980s, US acid rain politics were mainly reactive. Acid rain moved onto the White House's political agenda when the Canadian government pushed for a bilateral accord or when congressional initiatives on acid rain threatened to affect the interests of industry.

NAPAP did not occupy the center of the administration's attention. In the years leading up to the Interim Assessment, White House staff occasionally referred to NAPAP as the "multi-million dollar research program."[63] When the White House needed information on acid rain, it did not turn to the federal acid rain program but set up its own committees and task forces. The DPC, the policy forum for acid rain, had its own Working Group on Energy, Natural Resources and the Environment. The Cabinet Council on Natural Resources and Environment Working Group on Acid Rain Policy, established in July 1983,[64] had some overlap with NAPAP's Interagency Task Force. In April 1987, the DPC established an interagency group to determine costs and economic benefits of NO_x and SO_2 control programs.[65]

In January 1983, the administration discussed a reform of the Acid Precipitation Task Force, NAPAP's highest organizational body. A. Alan Hill, the new CEQ administrator, pushed for making CEQ the lead agency in NAPAP: "CEQ should be formally designated as the lead agency for acid rain with responsibility to serve as the Executive Secretary of the Acid Precipitation Task Force and as the Administration's overall acid rain policy implementor and spokesman." Office of Management and Budget and EPA administrator Anne Gorsuch and senior staff at the EPA concurred. Hill justified this proposal by noting: "There is no one Administration appointed official currently overseeing the activities of these lead agencies. Coordination of the research program, negotiation process and policy development is needed."[66] Edwin Meese, counselor to the president, also supported this position. He saw a "growing need for overall policy coordination . . . especially with respect to the Acid Precipitation Task Force." He determined in January 1983 that the CEQ should be the lead coordinating agency for acid precipitation (as it was designated in 1980), with the explanation that it was the only agency in the task force not directly involved in research.[67] Meese's remark supports the view that the agencies were primarily interested in pursuing their individual research projects with NAPAP funding and not in coordinating their research.

THE INTERIM ASSESSMENT: WHAT WAS THE CONTROVERSY ABOUT?

J. Laurence Kulp and the Interim Assessment

In 1985, Bernabo announced that he would step down as NAPAP director in order to work on new projects. The Joint Chairs Council appointed J. Laurence Kulp as director.[68] Kulp was a geochemist who had received his PhD in physical chemistry from Princeton in 1945. Between 1947 and 1965, he had been a professor of geochemistry at Columbia. His main area of research was radiometric dating and radioactive fallout. In 1976, Kulp became vice president for research and development at the Weyerhaeuser Company. The evidence suggests that William Ruckelshaus, then head of EPA, expected senior manager Kulp to tie the multiple NAPAP-funded research projects together and strengthen the focus of the program in order to get the Integrated Assessments under way. Mahoney said:

> Bill [Ruckelshaus] thought that Larry would be a perfect fit for this. And it was like "Larry is older and kind of a senior manager while Chris [Bernabo] is younger and very much a scientist. . . ." Bill's view was getting somebody who was ready to really lay down the law, this is really what we need to do and get things done.[69]

When Kulp became director, the organizational structure of NAPAP changed (see figure 2.3). The Assessment Task Group was dissolved, and its duties were transferred to Kulp's newly created role as the director of research. The intention behind the transfer of responsibility was to "achieve a higher level of overall program coordination and to assure that NAPAP assessment activities reflect a sufficiently broad range of policy concerns and issues."[70] According to the 1983 Annual Report, the Interim Assessment planned for 1985 should encompass four areas:

> First, an assessment of current damages attributable to acid deposition, second, an uncertainty analysis of key scientific areas, especially emissions and atmospheric processes, third, the implications of uncertainty in these areas on policy alternatives, and, fourth, a description of the framework for the Integrated Assessment methodology that should be the basis of the Integrated Assessments that were scheduled for 1987 and 1989.[71]

However, the Interim Assessment that NAPAP actually released, after a two-year delay, looked very different.

Instead of addressing a "broad range of policy concerns and issues," the Interim Assessment focused on NAPAP's research program and in particular on the uncertainties related to acid rain. When Kulp came into office in September 1985, he decided to significantly rework the draft of the assessment written under Bernabo. He declared that the report needed to be revised in order "to make it substantive rather than speculative."[72] NAPAP repeatedly moved back the final draft's release date, finally publishing it on September 17, 1987.

At the request of John D. Dingell, then chairman of the Subcommittee on Oversight and Investigations, Committee on Energy and Commerce of the House of Representatives, the General Accounting Office reviewed NAPAP a few months before the Interim Assessment was released. The review, *Acid Rain: Delays and Management Changes in the Federal Research Program*, published in May 1987, offered some insights into the changes in management when Kulp took over:

> The scope of these proposed assessments has been scaled back over the years. For example, NAPAP's plans to include economic analyses in its first assessment document have been revised, and it is not clear how much, if any, economic work will be included in future NAPAP assessments. Also, NAPAP officials told us they are uncertain whether NAPAP will be able to meet its long stated goal of producing by 1990 an "integrated" assessment that estimates the costs and benefits of various control options.[73]

Just how much of this delay in producing planned parts of the assessment can be attributed to Kulp is unclear.

The General Accounting Office (GAO, now Government Accountability Office) review suggested that the power of the agencies severely hampered Kulp's influence. The report stated that Kulp had "no real authority over the agency representatives participating in NAPAP."[74] In particular the DOE had a substantial impact, evidenced by NAPAP's extensive and optimistic discussion of clean energy technology.[75] The DOE further influenced the Interim Assessment by introducing the controversial statement that acid rain affected not "most" or "many" but only "some" lakes.[76] However, Kulp's personal view on acid rain concurred with the DOE perspective. His statements about acid rain after he left the program demonstrate that he believed acid rain to be a relatively small problem that would eventually be fixed through technological innovations. In his view, the CAAA of 1990 were expensive and not warranted given the small economic damages caused by acid rain.[77]

Kulp tried to influence NAPAP's research program, but he was not able to move NAPAP research in directions that the agencies and agency researchers did not want to go. The GAO report noted that he had "minimal control over the budget, which limits his ability to shape the research program."[78] *Science* reported in 1987 that Kulp worked individually with scientists on the chapters of the Interim Assessment and that he tried to impose his views on specific topics: for example, "Larry is pushing his own beliefs on oxidants" (referring to chemicals like ozone that are produced by various pollutant precursor emissions, including nitrogen oxides).[79] But NAPAP researchers strongly resisted Kulp's efforts to include ozone in NAPAP as a pollutant that likely causes damages to forests. This was partly a question of what pollutants should be included or excluded in an assessment of acid rain and partly a question of the interests of the NAPAP researchers. For example, meteorologist Bruce Hicks stated: "Kulp wanted to include ozone, but the chemists wanted to have an independent ozone assessment."[80] By including ground-level ozone, Kulp would have redefined and greatly broadened the problem. But he would also have shifted the regulatory burden away from the sulfur dioxide emissions of electric power plants.

The Interim Assessment and Its Critique

The NAPAP Interim Assessment was published in 1987 in four volumes: an executive summary, *Emissions and Controls*, *Atmospheric Processes and Deposition*, and *Effects of Acidic Deposition*, the latter comprising five chapters dealing with the effects on agricultural crops, forests, aquatic systems, materials, and human health and visibility.

NAPAP released its Interim Assessment at a time when the debate over whether the United States should impose regulations to combat the effects of acid deposition had been going on for years. In 1985, a review of several acid rain assessments by seven experts in acid rain research (Charles T. Driscoll, James N. Galloway, James F. Horning, Gene E. Likens, Michael Oppenheimer, Kenneth A. Rahn, and David W. Schindler) had concluded that there was a broad scientific consensus on the damaging effects of acid deposition in the United States.[81] In 1986, President Reagan and Prime Minister Mulroney had endorsed the report by their Special Envoys on Acid Rain, which accepted that acid rain was a serious problem. But there was still no domestic regulation, and by 1987 the American-Canadian negotiations had not made progress on a bilateral accord.

The Interim Assessment sparked a debate about scientific integrity that reached well beyond the realm of academia and led to severe tensions in US-Canadian relations. As Congressman James Scheuer noted in his opening statement in a special hearing on NAPAP before the House Subcommittee on Natural Resources, Agriculture Research, and Environment in 1988, the Interim Assessment raised "a storm of controversy."[82] The Canadian environment minister, Tom McMillan, attacked the assessment as "voodoo science" and stated that the document was "awkwardly out of step . . . with the prevailing scientific judgment on the subject."[83] The Canadian acid rain program, the Canadian Federal/Provincial Research and Monitoring Coordinating Committee (RMCC), which included Canadian government scientists, criticized the assessment for its "misleading, flawed, and incomplete conclusions."[84] Gene Likens characterized the report as "badly misrepresent[ing] the general scientific understanding about air pollution and acid deposition."[85] The National Research Defense Council called the Interim Assessment "nothing more than political propaganda."[86]

The trigger for the controversy lay in the Interim Assessment's conclusion that the effects of acid rain were not serious:

> Available observations and current theory suggest that there will not be an abrupt change in aquatic systems, crops, or forests at present levels of air pollution. Some lakes and streams in sensitive regions appear to have been acidified by atmospheric deposition at some point in the last 50 years. Available data suggest that most watersheds in the glaciated Northeast are at or near steady state with respect to sulfur deposition and that further significant surface water acidification is unlikely to occur rapidly at current deposition levels. . . . At current levels of acidic deposition, short-term direct foliar effects on crops or healthy forests are unlikely.[87]

For the administration, this reaction came as no great surprise. Kulp was supposed to testify in Congress about the forthcoming Interim Assessment's conclusions in early July 1987.[88] Ralph Bledsoe, special assistant to the president, expected that "some of his testimony will likely trigger headlines and various types of responses from environmental interest groups." He wrote to Nancy Risque, the cabinet secretary, that "we should be thinking about the White House response, if any, that may be required" after the release.[89] On August 5, EPA administrator Thomas told the DPC that the Interim Assessment "should generate much dialogue, as some congressional views of the science are likely to differ with the report."[90] Thomas confirmed that the NAPAP Interim Assessment, to be released the following month, would be a consensus document by all six member agencies of the Joint Chairs Council.[91]

Kulp presented the Interim Assessment in the DPC five days after its September 17 release and called it a "scientifically sound effort by the US, including Canadian participation." The minutes of this meeting continued that "he felt the criticisms appearing in the press are ill-founded and not generally supported throughout the scientific community." The DPC members found the briefing "very informative" and "said they would expect the conclusions contained in the report to be helpful in developing future acid rain and related policies."[92]

The administration downplayed the early critique by remarking that "some environmental groups and some people in Canada . . . felt it [the Interim Assessment] minimized the harmful effects of acid rain."[93] But instead of declining after a few days, the criticism became stronger. In December 1987, the RMCC released a comprehensive critique, and in January 1988 the Canadian environment minister wrote to the EPA expressing his disagreement with NAPAP's conclusions.[94]

As the Canadian government took initiative, the Reagan administration started to pay attention to its acid rain program. The White House wanted to make sure that the federal acid rain program got out of the news. Kulp quit as director. Unlike the procedure when Kulp became director, the new director, Mahoney, not only was interviewed by the Joint Chairs Council but also met with Bledsoe:

> He [Bledsoe] met with me and we talked for half an hour or 45 minutes, but I remember [in] particular what he said before we finished was "the best thing you can do for us by far is get this thing out of the newspapers and off the television. . . . The administration is being hurt politically with what is going on now."[95]

This supports the view that the Reagan administration's acid rain policy was mainly defensive.

Different Layers of the NAPAP Controversy

The debate over the NAPAP Interim Assessment involved different actors (in particular scientists and policy makers in both the United States and Canada and NAPAP officials) and different though related layers: first, a scientific layer (What were the scientific findings?); second, a scope layer (What did the assessment include and exclude? How was the phenomenon defined?); third, a methodology layer (Which sampling techniques were used? What kinds of evidence were brought up to support the conclusions?); fourth, a procedural layer (How was the executive summary compiled from the findings of the reports? How were the data interpreted? Was there peer review?); and fifth, a policy layer (Did the assessment meet its goal in terms of providing knowledge relevant for policy making? Was there a hidden goal, e.g., stalling political action?).

The following section gives a short overview of the main issues generating criticism published in newspapers and voiced in letters and congressional hearings in the weeks and months after the release. This critique came mainly from university scientists in the United States and from the Canadian government.

Historical Patterns of Acid Deposition

Knowledge of historical patterns of acid deposition is crucial for understanding causes and effects. However, data were scattered, and assessments concluded that it was difficult or impossible to establish reliable historical trends.[96] There was a high degree of consensus that SO_2 emissions had increased in the twentieth century,[97] but the joint report of NAPAP and the RMCC[98] and the NAPAP Integrated Assessment both pointed to a declining trend in recent years.

NAPAP's data on the historical patterns of emissions were not controversial, but the way in which the executive summary presented the data was criticized. The executive summary noted that "after peaks during World War II (24 million metric tons/yr) and the early 1970s (about 30 million metric tons/yr), sulfur dioxide emissions in 1985 (21 million metric tons) were about the same as in the late 1920's."[99] In fact, as a result of the CAA of 1970, emissions declined during the 1970s. The critique focused on both the presentation and the interpretation of the data: even if emissions had declined since the early 1970s, they were still high. McMillan illustrated the consequences of the way in which historical trends were presented: "The Executive Summary could have stated that US SO_2 emissions have increased by more than 200 percent since 1900."[100]

The Extent and Definition of Impacts on Aquatic Ecosystems

Damage to aquatic ecosystems caused by lake acidification was one of the best-documented effects of acid rain and had been investigated in the United States and Canada since the early 1970s.[101] The evaluation of aquatic effects in the executive summary of the Interim Assessment is an illustrative example of how phrasing suggested that the effects of acid deposition were of minor importance: "Some lakes and streams in sensitive regions appear to have been acidified by atmospheric deposition at some point in the last 50 years."[102] The same basic claim can also be found in the research chapter on aquatic effects in the Interim Assessment ("Acidic deposition has contributed to the acidity of some northeastern and midwestern acidic lakes") and in the Integrated Assessment ("Acid deposition has caused some surface waters to become acidic in the United States").[103] In contrast to the executive summary, however, these two reports discussed the extent of acidification in more detail and did not include qualifying words such as "appear to" and "at some point in the last 50 years."

Reviews of the Interim Assessment widely rejected the statement that acid deposition affected only some American lakes. In particular, scientists contested the reference value of pH 5 for an acidic lake, used in both the NAPAP executive summary and the aquatic effects chapter of the Interim Assessment, stating that scientists had observed damages at pH values above 5.6 and even above pH 6.[104] In its document *A Critique of the US National Acid Precipitation Assessment Program's Interim Assessment Report*, the RMCC asserted that not "some" but "numerous" lakes had been acidified.[105]

Why did NAPAP and the RMCC come to different conclusions? There were several differences between the RMCC and NAPAP assessments that accounted for the fact that the extent of aquatic damage appeared to be much smaller in the NAPAP reports. While the RMCC set the pH reference value at 6, the NAPAP Interim Assessment chose pH 5 as the reference value. One step down on the pH scale means a tenfold increase in acidity and significantly decreases the number of lakes recognized as acidified. In its response to its critics, NAPAP noted that validating the "general statement that serious damage begins to occur at pH 6.0" required detailed assessments and that "the Interim Assessment states that biological responses can occur at pH levels above 5.5."[106] The NAPAP authors argued that their aim was "to inventory 'acidic' lakes; i.e. reflecting the status of lakes rather than any implications about the source of acidity, the date of its origin, or the presence or absence of biological change at other pH levels." However, the choice of pH 5 was in no way a value-free choice to "inventory" acidic lakes. The disagreement was about the pH level at which "damage" occurs, in contrast to "changes" that might not qualify as "damage." This became clear when the NAPAP authors noted in their response that "it should not be assumed that all variations in aquatic chemistry and biology are due to acid deposition, or that chemical and biological variability is synonym[ou]s with damage."[107]

Another difference was in the presentation of the numbers. The RMCC gave the total number of affected lakes—which was a large number—while NAPAP gave the percentage of lake area affected—which was a small number. Likens has pointed out that the adequate measure for the extent of lake acidification is not the percentage of acidified lake area compared to the overall lake area in the United States but the impact in "*affected* areas. Stating a fraction on the basis of the entire world or the universe would even further minimize the impact!"[108] The RMCC further criticized NAPAP for only focusing on US lakes and not on the effects of US emissions on Canadian lakes (see table 2.1).

The NAPAP executive summary concluded that "most watersheds in the glaciated Northeast are at or near steady state with respect to sulfur deposition" (i.e., their acidification status was not changing) and that further acidification "is unlikely to occur at deposition levels."[109] McMillan, the RMCC, and Likens responded that this hypothesis was not supported by sufficient evidence.[110] The RMCC noted that many studies indicated that "sensitive lakes in the Northeastern part of the continent are not in steady state with respect to current levels of emission/deposition and that chemical and biological deterioration is continuing."[111]

Another point of criticism was that NAPAP concentrated on absolute pH values rather than changes in pH values and that the assessment did not include a discussion of the sensitivity of lakes to acidification.[112] While this was true for the executive summary, it did not apply to the National Surface Water Survey or the aquatic effects chapter. Both the aquatic effects chapter and the NSWS gave results for different pH reference values (5.0, 5.5, and 6.0) and investigated the sensitivity of lakes (i.e., their acid-neutralizing capacity). The aquatic effects chapter, however, defined an "acidic lake" as a lake with a pH of 5.0 or lower, as did a report from the US Office of Technology Assessment.[113]

Different conclusions as to whether "some" or "numerous" lakes had been acidified resulted from decisions about which reference value (pH 5.0, 5.5, or 6.0) should be used and which reference framework should be chosen (only sensitive areas, all US lakes, or all North American lakes) and thus from judgments about what qualified as a substantial damage, which in turn reflected the bureaucratic and political priorities of agencies.

Terrestrial Effects—Forest Decline

Scientists first noticed forest decline in East and Central Europe in the late 1970s, and it became a broadly discussed topic in the early 1980s. While

TABLE 2.1. NAPAP and RMCC assessments of the effects on aquatic ecosystems (emphasis added).

Assessment	*Aquatic Effects*
NAPAP, *NAPAP Interim Assessment*	"*Some* lakes and streams in sensitive regions *appear* to have been acidified by atmospheric deposition *at some point* in the last 50 years." (1:I-8) "Lakes and streams that appear to have been acidified, at least in part, by atmospheric deposition represent a *small fraction* of the surface water in the United States." (1:I-9)
RMCC, *A Critique of the US National Acid Precipitation Assessment Program's Interim Assessment Report*	"The omission of Canadian modelling and survey results leads to an *under-estimation of the severity* of acidification of surface waters. For example, there are 700,000 (approximately) lakes in Eastern Canada below the 52nd parallel. Canadian surveys indicate that, of these, 16% in Quebec (72,000) and 23% in Ontario (29,000) have *pH's below 6*. Compare this with the total population of 33,500 lakes in the Adirondack region (11,000), Florida (9,000), and Western U.S. (13,500), combined." (24)
NAPAP, *NAPAP Analysis of the Canadian RMCC Critique of the NAPAP Interim Assessment*	"NAPAP recognizes that acidic deposition has contributed to the acidity of *some* Canadian lakes. . . . The Canadian lake estimates (when expressed as a percentage of the total population of lakes) are generally similar to results from northeastern regions in the US Eastern Lake Survey. The percentage of lakes with *pH less than 5.0* ranges from 1 to 10 percent in subregions of both eastern United States and eastern Canada, with an average of about 2 percent in eastern Canada." (14)
NAPAP, *1990 Integrated Assessment Report*	"Within acid-sensitive regions of the United States, 4% of the lakes and 8% of the streams . . . are chronically acidic. Overall 263km² (2%) of the 12,000km² of lake area in the NSWS was acidic and 1,310km² (11%) had *ANC ≤ 50 meq/l*. . . . Acid deposition has caused *some* surface waters to become acidic in the United States." (11)

most scientists were convinced that air pollution played an important role in forest damage and decline, it was not clear whether acid deposition was involved or, if it were, how acid deposition related to other potential damaging effects caused by oxidants like ozone or hydrogen peroxide.

There were several competing hypotheses to explain forest decline due to acidic deposition. A review article from 1985 distinguished among aluminum toxicity theories, gaseous pollutant (principally ozone) theories, foliar nutrient deficiency theories, the general stress hypothesis, and the nitrogen hypothesis.[114] The increasing importance of the issue in the United States was evidenced by a substantial increase of funding: in 1984, NAPAP's forest research budget was less than $1 million; this amount increased to over $10 million in 1985, and to over $18 million in 1986 before decreasing slightly in 1987 to $16 million.[115]

During the 1980s, the state of knowledge improved and most assessments concluded that the evidence showed effects from acid deposition on terrestrial ecosystems (see table 2.2). Only two[116] of the eight assessment reports from the years 1982 to 1990 that discussed this issue concluded that no evidence for these effects existed; one of these was NAPAP's Interim Assessment. The joint report by NAPAP and RMCC included two different positions. Canadian and US scientists referred to the same studies on seedlings but came to different conclusions. The Canadian position was that treatment with rain that had a pH ≤ 4.6 induced "significant growth reduction and morphological changes," while the US interpretation was that "except at pH of 2.6 there were no significant effects on cumulative germination, survival or above ground biomass. Effects on some morphological characteristics were ambiguous at pH 3.6 and 4.6."[117]

NAPAP's executive summary stated that "negligible effects" of acid deposition on forests had been found and that it was unlikely that regional sulfur dioxide was causing damage to forests and crops.[118] Evidence for this conclusion came from experiments in which seedlings were exposed to simulated acid rain. As no damage was observed down to pH 3.5, and, allegedly, because rainfall rarely fell below pH 4.1, both the executive summary and the research report chapter of the Interim Assessment concluded that damage was unlikely.[119] These conclusions were broadly criticized; most of the critics pointed out that seedling experiments have only limited significance for statements about real forests, that the pH of rainfall is often below pH 4.1, and that the effects of combined stresses are important. Scientists argued that—contrary to NAPAP's conclusions—abrupt changes to forests do occur, and acid rain can cause damage to unhealthy trees or where trees are exposed to various natural and pollutant stresses.[120]

While the reports agreed that there was substantial uncertainty related to forest effects, they differed in the ways they framed this uncertainty and weighed available evidence. The discussion about damage to forests is in some respects similar to the discussion of aquatic effects. For aquatic effects, scientists disagreed about reference values and thus on the definition of substantial damage; for forest effects they disagreed about approaches to measure environmental damage. Today there is a consensus that acid rain contributes to some cases of forest decline but is only one of many causes.[121]

Atmospheric Processes

Differing views of the science of atmospheric processes determined how the reports described the formation and transport of acidic substances in the

TABLE 2.2. NAPAP and RMCC assessments of forest effects (emphasis added).

Assessment	*Forest Effects*
NAPAP, *Annual Report to the President and Congress (1983)*	"*Some forests* in North America and Europe are showing environmental stress, the *causes of which are a topic of scientific debate.*" (2)
NAPAP, *Annual Report, 1985*	"As yet, *no scientific consensus* exists in the United States to link acid deposition with apparent changes in forest condition at the regional level." (79)
RMCC, *Assessment of the State of Knowledge on the Long-Range Transport of Air Pollutants and Acid Deposition: Part 1: Executive Summary*	"The evidence to date suggests that air pollution, including acidic precipitation is *most likely* involved on a long-term basis in the decline of forests in several parts of the world. Direct association is not readily demonstrated since forest ecosystems are subject to *multi-stress* including air pollutants, insects, disease, adverse weather, and climate." (1-13)
NAPAP, *NAPAP Interim Assessment*	"At current levels of acidic deposition, short-term direct foliar effects on crops or healthy forests are *unlikely*. Acid deposition may have a cumulative effect on trees growing on certain low-nutrient soil, but this effect is expected to be *gradual* and has not been reported in the United States. It is *unlikely* that regional sulfur concentrations are causing damage to crops or forests." (1:I-8)
RMCC, *A Critique of the US National Acid Precipitation Assessment Program's Interim Assessment Report*	"There is no mention [in the *NAPAP Interim Assessment*] of the unexplained dieback of low elevation hardwood forests in the northeastern USA and southeastern Canada in recent years, particularly the extensive dieback of sugar maple in Quebec. This dieback is located in the region of heaviest deposition and may be a phenomenon in which acid deposition *plays a major role.*" (4) "The [*NAPAP Interim Assessment*] report dismisses the *present damage in the forest environment.*" (21)
NAPAP, *NAPAP Analysis of the Canadian RMCC Critique of the NAPAP Interim Assessment*	"The forestry section of the Executive Summary [*NAPAP Interim Assessment*] is process-oriented and does not specifically mention any of the unexplained forest diebacks in North America. . . . This section [chapter 7.2.6 of the Interim Assessment Research Reports] notes that soil effects from acid depositor resulting in sugar maple decline *might be occurring but have not been intensively examined or demonstrated* at this point." (20)
NAPAP, *1990 Integrated Assessment Report*	"There is *no evidence* of widespread forest damage from current ambient levels of acid rain (pH 4.0–5.0) in the United States. Localized areas of forest decline . . . do occur, however, as a result of the combined action of *multiple stress* factors. . . . Acid deposition *can increase* the total stress on the forest system." (45)

atmosphere as well as how they derived statements about relationships between emissions and deposition. The question of whether these relationships were linear, so that a reduction in emissions would lead to a proportionate reduction in deposition, or were basically nonlinear, so that a cut in emissions would not necessarily lead to a proportionate reduction in deposition, became a key question for science and policy making in the United States. The paragraph on source-receptor relationships in the NAPAP executive summary highlighted the various factors that accounted for nonlinear relationships and stated that although a reduction of deposition would occur when emissions were reduced, "the magnitude and extent of the reduction (of deposition) is uncertain."[122]

NAPAP's assessment of source-receptor relationships was one of four "serious omissions"[123] that the RMCC found in the report. In particular the RMCC criticized the executive summary for emphasizing the nonlinear relationships between emissions and wet deposition but not mentioning that dry deposition (which accounts for a considerable amount of overall deposition) behaves linearly. RMCC concluded that "Canadian studies have shown that reducing sulphur dioxide emissions will result in a proportional reduction in sulphur loading over large areas."[124] RMCC reviewers further stated that this chapter omitted work on cloud chemistry and raised doubts about whether the Regional Acid Deposition Model would be available in time to provide a state-of-the-art answer to the linearity question. RMCC called NAPAP's development of RADM an "obsession" that was "inexplicable in scientific terms"[125] and criticized it because "the extensive trajectory analysis work carried out by various workers, and the evidence for source-receptor relationships provided by this work, is omitted."[126]

Thus on the question about the nature of source-receptor relationships, the RMCC and NAPAP openly disagreed. In August 1988, NAPAP published a reply to the RMCC critique. It emphasized that "the analyses conducted to date provide a scientific foundation, sufficient to concern policy analysts, that nonlinearity exists" and that RADM had to be developed in order to evaluate the role that nonlinearity played.[127] However, while this statement made it seem as though NAPAP scientists had a clear position regarding the existence of nonlinearity, not all NAPAP scientists were convinced that nonlinearity played an important role. Even within NAPAP the debate was ongoing.

The Impact of the Interim Assessment Debate

The controversy over the Interim Assessment report quickly came to focus on Kulp. Congressman Scheuer remarked:

> This [the executive summary] was a personal achievement of Dr. Kulp's. He must have gone into the attic and written this 35-page report on his own because there was no peer review, there was no involvement of the Interagency Scientific Committee, as Congress certainly expected would happen. This was an individual creation of Dr. Kulp, and he quit within days after it was released.[128]

Mahoney and other NAPAP officials, however, stated that the executive summary was scientifically backed by the findings in NAPAP's research reports, and they defended the Interim Assessment in two responses to the critique.[129] When asked in a 1988 congressional hearing on NAPAP about "the reason for such a stark difference between the two items" (i.e., the scientific reports and the executive summary), Mahoney answered:

> I honestly believe that . . . the executive summary is not as distant from the underlying science as it is often characterized as. What I think I see . . . is a difference in emphasis, a difference in issues covered. It is hard to find something in the executive summary that you could call inaccurate based on what is in the document.[130]

In fact, almost all of the text in the executive summary was copied verbatim from the conclusions of the respective chapters of the Interim Assessment and did not contradict the results published there. But if the task of assessments is to paint the broad picture, it is precisely the difference in emphasis and the selection and presentation of the science that matter.

The Presentation of Science: Facts and Values in the Interim Assessment Debate

This difference in emphasis from other (Canadian, European, and US) acid rain assessments became particularly clear in the "major conclusions" section of the executive summary of NAPAP's Interim Assessment report. Conclusion I.3.1.2, "Probability of Changes in Effects at Current Emissions Levels," began with the statement "Available observations and current theory suggest that there will not be an abrupt change in aquatic systems, crops, or forests at present levels of air pollution."[131] Chapter 7 of the Interim Assessment (written by Kulp), in contrast, began with a statement about the importance of healthy forests and pointed out that the impact of air pollution on forests had been recognized on a local scale in the 1970s. However, it also stated that the impact of air pollution on a regional scale was unclear.[132] Conclusion I.1.3.3, "Relationship between Emissions and Acid Deposition," emphasized nonlinearity and uncertainty in the first few sentences, stating,

> In the northeastern United States, the formation of sulfuric acid in cloud water from sulfur dioxide and hydrogen peroxide appears to be limited by the availability of hydrogen peroxide in winter and perhaps in other seasons as well. This means that a reduction in the emissions of sulfur dioxide in the northeastern quadrant of the United States in winter is unlikely to result in a proportionate decrease in the formation and subsequent deposition of sulfuric acid over the northeastern United States. The magnitude and geographical extent of this so-called "nonlinearity" have not yet been evaluated.[133]

The matching chapter of the Interim Assessment also emphasized nonlinearity by highlighting the role of multiple sources, meteorological conditions, and the influence of local abundance of other pollutants that would account for the fact that "a change in sulfur dioxide emissions does not always lead to the proportional change in sulfur deposition at a receptor"[134] and by pointing out that in-cloud oxidation processes "can lead to nonlinear relation" between emissions and deposition because of the limitation of hydrogen peroxide (i.e., insufficient amounts of hydrogen peroxide, the oxidant that converts sulfur dioxide into sulfuric acid, may be present). While this chapter acknowledged that gas-phase chemistry was likely linear because the supply of oxidants was sufficient to convert the amounts of sulfur dioxide present, it noted that the relative importance of gas-phase versus aqueous-phase chemistry was not known.[135]

Conclusion I.1.3.4, "Estimated Economic Impacts of Acids and Ozone," began with the optimistic outlook that "at current deposition levels, there is no detectable effect of acidic deposition on crop yield; however, there may be a net fertilizer benefit from nitrogen deposition on the order of $100 million per year."[136] The conclusion of chapter 6 in volume 4 of the Interim Assessment determined more precisely the scope of this statement by stating, "The available information from scientific research has established no measurable and consistent crop yield response from the direct effects of simulated acid rain at ambient levels (pH 3.8–5.0)."[137] Conclusion I.1.3.5, "Multiple Stresses on Forests," emphasized in the first sentence that "all forests are subject to variable natural stresses, causing periodic growth suppression and/or visible injury" before admitting that "air pollution can add an additional element of stress."[138] The corresponding chapter in the Interim Assessment, written solely by Kulp, stated in its introduction that

> the possible existence of air pollution damage to forests at a longer regional scale is supported by the earlier point source cases, known agricultural effects . . . , and recent reports of widespread unexplained injury and/or reduced growth from some forest areas in Europe and North America.[139]

It also noted that results from seedling experiments "cannot be directly equated to the mature tree."[140]

The discussion of the critique of the Interim Assessment showed that much of the debate was about the interpretation and presentation of facts. Did the assessment conclude that acid rain was a serious problem and that actions for abatement were required? The Canadian environment minister pointed out that Canada's concerns "go beyond the question of science to the overall thrust and intent of the document. The Executive Summary portrays the problem of acid rain as neither serious nor in need of any intervention."[141]

The structure of the RMCC critique illustrated that Canadian government scientists were not only criticizing statements that could actually be found in the executive summary but were also addressing possible interpretations of these statements. The RMCC critique opened with a discussion of five "basic conclusions" of the NAPAP Interim Assessment that—according to the RMCC—were suggested by the executive summary:

I. The effects of acid rain are neither widespread nor serious.
II. There will be no abrupt changes in the effects of acid rain for the next several decades.
III. Emission levels of sulphur dioxide have been nearly constant since the 1920's, are currently stable, and will decrease substantially over the next three or four decades through the application of new technologies due to market forces.
IV. The effects of acid rain are less than were anticipated ten years ago.
V. Sufficient uncertainties remain to preclude whether abatement action is needed or the nature of this action.[142]

These conclusions were not explicitly made in the assessment but were derived from the statements made in the executive summary. The RMCC then responded to these suggested basic conclusions with five "facts": first, "The effects of acid rain are already widespread and serious"; second, "The effects of acid rain are worsening"; third, "The US SO_2 emission pattern has changed profoundly over the past decades and emission levels will not decrease in the foreseeable future"; fourth, "The acid rain problem is worse than it was anticipated ten years ago"; and fifth, "What is known about the nature, causes, and effects of acid rain is sufficient to design effective abatement programs."[143]

Both NAPAP assessments—the Interim Assessment and the Integrated Assessment—had a stronger inclination to assess a particular issue as uncertain or not documented than other US and Canadian assessments. From the beginning, NAPAP reports did not talk about substantial damage to aquatic ecosystems but stated that "some lakes" are affected (see table 2.1). In contrast, the NADP/CEQ report from 1978 described the aquatic effects in the United States as "catastrophic"[144] and the 1981 NRC report talked about a "severe degradation of many aquatic ecosystems."[145] Brysse et al. have argued that rather than alarmism, a more prevalent and perhaps systematic bias in environmental assessments is "erring on the side of least drama."[146] Within NAPAP, this tendency seems to have been particularly strong. This suggests that the bias toward "erring on the side of least drama" may be strengthened in instances when scientific knowledge is likely to be contested in political debates. Since agency scientists conducted NAPAP, their awareness of the political context of the assessment—namely the awareness of the position of their respective agencies—was strong. All of the interviewed agency scientists stated that they certainly represented their agency's position.

NAPAP did not take up the general critique made by the RMCC, stating instead that it would refrain from responding to "generalized critical statements," "policy issues," and the "suggested basic conclusions . . . since NAPAP does not agree that these reflect the statements in the Interim Assessment."[147] NAPAP thus restricted the discussion about the Interim Assessment to disagreement about facts and excluded the interpretation and meaning of this information from its discussion of the Canadian critique. The attempt of NAPAP scientists to stay away from value questions reflected their view of what was required for objectivity, and a clear attempt to demarcate science from values that was present throughout the program. However, policy relevance is one of the central characteristics of an assessment, and decisions about relevance are inherently normative and include value judgments.

NAPAP's position—as expressed in NAPAP documents—regarding its role in policy making changed significantly during the course of the program. In the very beginning, NAPAP's Interagency Task Force affirmed the congressional assignment "to identify actions to limit or ameliorate the harmful effects of acid precipitation"[148] by stating that NAPAP would develop policy recommendations.[149] However, in the following years, the same Interagency Task Group adhered to the view that NAPAP's task was to provide scientific and technical information (the research program established by the Acid Precipitation Act) as well as an evaluation of scientific knowledge for decision making (the assessment established by the Acid Precipitation Act) but that making suggestions for policy and control strategies was beyond NAPAP's scope. NAPAP documents insisted on a "rigorous separation of science and policy along with close coordination."[150]

In its 1983 Annual Report NAPAP addressed the "intensified" acid rain debate and the "emotionalism" surrounding the issue of acid rain and stated that while addressing "these urgent demands for information," it would maintain "the essential objectivity that is central to its scientifically-oriented, policy-neutral mandate."[151] "Policymakers," the report emphasized, "not researchers, must decide when scientific information is adequate for decisionmaking."[152] How much of this debate reflected genuine differences over what assessments should do and how much reflected agency exigencies and the political context is unclear.

But if the demarcation between science and policy was clear in NAPAP's statements about the roles of scientists and policy makers, in practice the line was blurred. In the Interim Assessment, for example, choices of reference values and geographic scopes were decisions that were bound up in judgments about how to define and determine environmental damage. The arrangement of facts in the executive summary blurred this line further by emphasizing what was not known instead of what was known and by presenting acid rain as a minor problem.

THE INTEGRATED ASSESSMENT: "PIECE-WISE NON-INTEGRATION"

The public controversy over the Interim Assessment had a noticeable impact on the program: "NAPAP was pretty much in flames," noted Jeremy Hales, an atmospheric scientist at the DOE's Pacific Northwest National Laboratory.[153] Because of the controversy, Kulp resigned shortly after the report's publication. Mahoney took on the directorship and introduced major changes in the program by setting up comprehensive assessment procedures. The scientists who had to prepare the final Integrated Assessment had to cope with four challenges: first, they had to create a framework for the Integrated Assessment; second, they had to integrate the research that had been done in various unrelated NAPAP-funded projects; third, they had to produce the report under high time pressure; and fourth, they felt insufficiently supported by the agency scientists, the agencies, and the administration. The following section analyzes the influence of these different challenges on the final report by focusing mainly on RADM, which became, as Malanchuk phrased it, "the weak link in everything."[154]

Challenge 1: Development of the Assessment Framework

The Integrated Assessment Framework

Until 1987, NAPAP did not have a functioning Integrated Assessment framework. Two Interim Assessments were originally scheduled for 1985 (Interim

Assessment) and 1987 (Interim Integrated Assessment), but due to the two-year delay of the Interim Assessment, the Interim Integrated Assessment report was never written. Mahoney spent his first month in office setting up a framework for the Integrated Assessment, starting almost from scratch. Neither NAPAP managers nor the scientists involved in the program had a common understanding of what the assessment should do or even of what an assessment was in the first place. In the first five years, NAPAP emphasized economic assessments. In Mahoney's account of what an assessment was, he emphasized that science had to be used for scenarios and for projections that lay out clearly what the different outcomes of different choices would be. This included an economic dimension but it was broader than just this:

> One of the big troubles with assessment work like this is often you do these individual studies—and you can say that's what Chris [Bernabo] was doing too—is like "We need better studies in all these areas," but if that's all you do with it, then this is not really an assessment. An assessment should be a decision analysis process at the end of the day. You advance your underlying scientific understanding, but then you use your science also to make projections, and of course projections have uncertainties, so you have to factor those into the consideration, but the big issue . . . at the end of the day is how would you examine possible future scenarios and say what would be better or worse with a different scenario. I remember I often talked with staff and the key agency representatives about the specific question of "Does it make any difference if you impose one kind of control?" Well, you take the policy things that actually happened, in terms of reducing the emissions of SO_2 from power plants, but not addressing other sources, does it make any difference if you make it for those compared to something else or would you get about the same results? . . . What is the metric for effects anyway? . . . Can we monetize at the end of the day? What is the cost of the damage? In a perfect science world you want to put everything on the same metric, which might be dollars.[155]

Although NAPAP made efforts toward an economic analysis, the ambitious plans from the beginning of the program were only very modestly realized.

WAZECK: How strong was cost-benefit analysis in the final assessment?
MAHONEY: Slightly better than weak. "Weak" would really be quite weak. It was a little bit more than just that, but not much.[156]

One of the first things Mahoney did after beginning his directorship was to send memos to the hundreds of scientists involved in NAPAP explaining that they would make a fresh start and that the first thing they would need was a common understanding of the assessment:

> The idea was, you can't just assume that everybody knows what is in an assessment, because everybody can have some view on this. . . . Central to the whole plan of the assessment plan itself was agreeing on the questions that would be addressed in the assessment. . . . When I got this group together, the first morning I was outlining that there is all this good work and that it is my responsibility to know this work in the various areas well enough to be pretty up on the sciences, but the question was: How does this fit together?[157]

Mahoney describes here the key problems for NAPAP's Integrated Assessment: research had been conducted in an extensive number of fields and on a high variety of subjects, the major NAPAP-funded research projects had each often been in the hands of one agency, there was a substantial lack of interaction and communication between the different research endeavors, and consequently, integration turned out to be difficult when the program was approaching its end (see the section "Challenge 2: The Integration of Research" below).

In October 1988, NAPAP issued a 130-page draft of the *Plan and Schedule for NAPAP Assessment Reports* that included the guidelines for the assessment on which all agencies had agreed:

- It must be *credible*, both to scientific reviewers and to the users of the assessment information. Therefore, the development of the assessment must be open to public review, and the underlying information must be fully peer reviewed.
- The assessment must be *comprehensive*, by examining the entire range of plausible causes, effects and control approaches.
- It must be *critical*, endorsing hypotheses that are supported by scientific research and rejecting unsubstantiated hypotheses.
- It must define scientific *confidence* levels for its findings, reflecting the unavoidable scientific uncertainties in complex environmental research.
- It must be *comparative*, evaluating alternative strategies to define a range of future effects likely to result under various scenarios.[158]

The assessment was structured around five questions:

> What are the effects of concern and what are the relationships between acidic deposition/air pollutant concentrations and effects? . . . What is the relationship

> between emissions and acidic deposition/air pollutant concentration? . . . What is the sensitivity to change? . . . What are projections of future conditions based on illustrative future scenarios? . . . What insights can be drawn from comparative evaluations of illustrative future scenarios?[159]

These questions partly overlapped the earlier assessment questions, but the last two, addressing the use of scenarios, were new.

NAPAP's 1990 final report consisted of two parts: The first part was the 27 State of Science/Technology (SOS/T) reports in four volumes (volume 1: *Emissions, Atmospheric Processes, Deposition*; volume 2: *Aquatic Processes and Effects*; volume 3: *Terrestrial, Materials, Health, and Visibility Effects*; volume 4: *Control Technologies, Future Emissions, and Effects Valuation*), altogether totaling more than 6,000 pages. The SOS/T reports were "comprehensive analyses and discussions of relevant technical information prepared for specialist readers." The second part was the Integrated Assessment report, "a structured compilation of policy-relevant technical information presented in a form suitable to assist policy makers and the public in evaluating the key questions concerning acidic deposition causes, effects, and control strategies."[160]

The Integrated Assessment concentrated on the development of future scenarios, in particular future emission scenarios. These scenarios were meant to provide the inputs for the linked models that would then provide results for comparison. (Why this integration largely failed is discussed below.) Scenarios considered mainly reflected different SO_2 emission changes to a 1980 baseline. Examining SO_2 reduction scenarios of 12, 10, and 8 million tons, the report suggested that, when compared to the 10-million-ton reduction, the benefits would be only "marginally greater" with the 12-million-ton reduction and only "marginally smaller" with the 8-million-ton reduction. Comparing scenarios embodying a 10-year versus 30-year delay of policy implementation showed that "timing of changes in average deposition and air quality will generally coincide with the timing of changes in emissions."[161] Half or more of sulfur deposition came from sources not more than 500 km away from sensitive ecosystems, and reducing emissions from a mix of sources rather than from industrial sources alone would have "little effect" on large-scale deposition patterns.[162]

Dealing with Uncertainty

For the Integrated Assessment, NAPAP required a system for dealing with uncertainty. For most of its life, NAPAP had no formal way of handling uncertainty. In 1988, the "star system," a system occasionally referred to as the "restaurant rating guide," was devised.[163] This system expressed the relative confidence of experts in each finding with stars: zero stars: no basis for an answer; one star: some information but major uncertainties and knowledge gaps; two stars: adequate information but generally large and ill-defined uncertainties; three stars: ample information with well-defined but sometimes large confidence intervals; four stars: substantial amount of consistent information. The task groups were responsible for assigning stars to statements, and there was pressure to achieve a star distribution that would "look good"[164] in the final report.

Malanchuk stated that the star system's qualitative way of dealing with uncertainty was "probably the best that we could do," but, he admitted,

> it wasn't really very satisfying at the end, because then you get into some of the task group[s], and you would have these discussions about "Well, we spent 10 years, and you don't have a single four-star conclusion? How many one-star conclusions do you have?" . . . In order to get the highest level of confidence, you had to have some statement that was completely stupid. But you had to have a mix of—it had to look good.[165]

Indeed, there were star ratings in the Integrated Assessment that expressed a high level of confidence in the *uncertainty* rather than in the occurrence of an event. For example, the statement "Although [emissions] trading provisions are projected to result in reduced costs, the impact with regard to what states or regions will be buying or selling emissions reductions is highly uncertain"[166] was ranked as having a high level of confidence (four stars). *Science* reported skepticism about the star system among NAPAP researchers:

> Researchers were asked to estimate (using expert judgment) the quality of the information based on a scale of zero to four stars—as if they were reviewing restaurants. "Is a famous scientist's estimate of one star worth the same as a young guy's three stars?" asks one obviously skeptical NAPAP participant. "What if one expert ranks a hypothesis with two stars, but another ranks the opposite hypothesis with two stars? Should you rank the net at zero? How do you combine a one-star estimation and a four-star? Do they honestly expect Congress to think this means anything?"[167]

The approach to uncertainty was refined in later assessments, including IPCC's, but NAPAP did not achieve a comprehensive uncertainty analysis for the Integrated Assessment.

Challenge 2: The Integration of Research

Models and the Integrated Assessment

"Going for Broke on a Mega Model," read a 1991 headline in *Science*; the accompanying article went on to say that "in a sense . . . RADM epitomizes the best and the worst of NAPAP."[168] RADM was praised for its accomplishment in modeling of atmospheric processes using an approach "at the forefront of scientific capabilities," but it was the source of major difficulties when adopted for the purposes of the Integrated Assessment. RADM was the NAPAP-funded research project in which the structural tension between research orientation and assessment orientation became most apparent.[169]

RADM converted alternative emissions scenarios into geographic and temporal patterns of acidic deposition. These patterns became inputs to models that generated estimates of acidification impacts on lakes, forests, materials like building surfaces and infrastructure, visibility, and health. These estimates would be used as inputs for economic models projecting the economic costs and benefits of different scenarios (see figure 2.4).[170]

The development of RADM started in 1983 in order to answer a question at the center of the US acid rain debate in the 1980s: What was the relationship between sources of emissions and receptors of deposition? Source-receptor relationships were one of the most controversial issues, among both scientists, for whom it was a scientifically highly complex issue,[171] and policy makers, for whom a key question was whether regulations should be regionally applied or source specific.

In a 1985 congressional hearing on NAPAP, EPA administrator Lee M. Thomas described atmospheric processes as one of the "major uncertainties" related to acid rain and emphasized the importance of improved knowledge about source-receptor relationships:

> What are the current patterns of emissions and deposition, not just wet deposition but, as indicated, dry deposition in this country? [This is] a basic question that needs answers in this regard. How would deposition patterns change with changes in emission patterns? We're talking about long-term, long-range emissions. We're talking about long-range effects. Included in that is atmospheric and chemical transformations, major uncertainties and questions associated with deposition patterns and with emissions patterns associated with and related to that.[172]

Would a reduction in emissions lead to a proportionate reduction in deposition? Where should emissions be reduced in order to decrease deposition

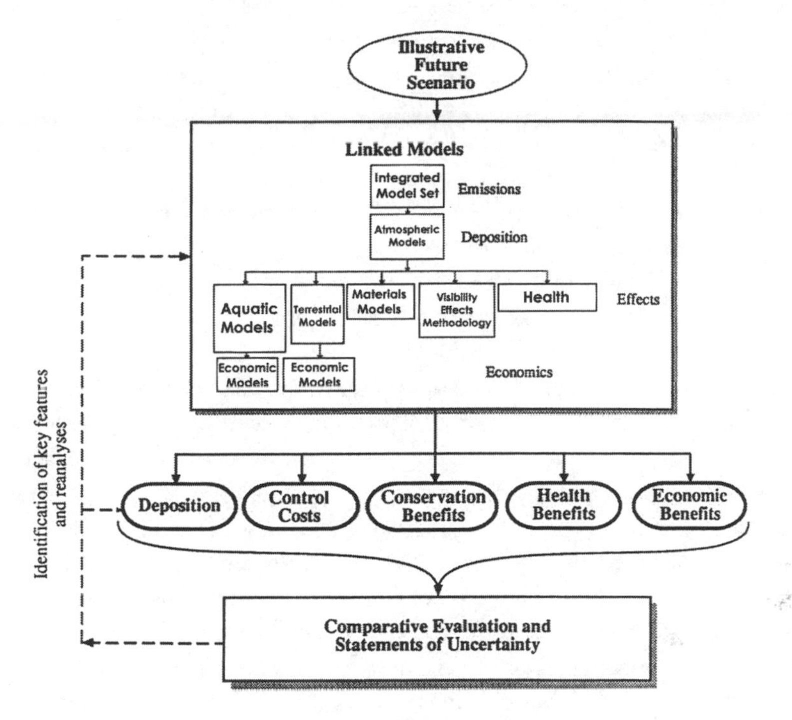

FIGURE 2.4 NAPAP's Integrated Assessment Framework (adapted from National Acid Precipitation Assessment Program, *Models Planned For Use in the NAPAP Integrated Assessment*, figs 2-2 and 2-3).

at specific sensitive sites? Would it make a difference whether 10 or 12 tons of sulfur dioxide were taken out of the air? These specific information needs dominated the policy debate in the mid-1980s, but (as shown below) in the end, they did not play a role in designing the CAAA.

The Science of Source-Receptor Relationships

The damaging effects of acidic pollution near sources such as smelters had been known for a long time,[173] and relationships between emission sources and effects in the nearby environment were relatively easy to establish. However, it proved much more difficult to establish source-receptor relationships over long distances in the atmosphere, where pollutants were subjected to complex transport and chemical transformation processes. As pollutants emitted from tall smokestacks remain for a long time in the atmosphere, chemical reactions can take place over a longer time period and in faraway regions.[174] In addition, reactions are complex; not only is the formation of sulfuric and nitric acid related to the presence of SO_2 and NO_x in the atmosphere, but the existence of NO_x also affects the formation of sulfuric acid. Furthermore, the formation of both sulfuric and nitric acid is related to the presence of ozone and hydrocarbons in the atmosphere. The amount of sulfuric acid that resulted from a certain amount of SO_2 in the atmosphere could thus be higher or lower, depending on the amount of NO_x and other pollutants in the atmosphere.[175]

Meteorological processes were a further complication. It had been demonstrated that source-receptor relationships were linear with regard to SO_2-to-sulfate gas-phase reactions. However, scientists had noticed in the late 1970s that the aqueous reactions, for instance when SO_2 reacts in clouds or rain, involved nonlinear effects, as did dry deposition of SO_2 on surfaces.[176]

While the early European studies on long-range transport and acid rain[177] did not specifically address the question of linearity versus nonlinearity, they implicitly assumed that nonlinearities had no significant overall effect on acid deposition. In the technical meetings of the US-Canadian memorandum of intent working groups in the early 1980s, it had been acknowledged that chemical processes leading to acid deposition were likely nonlinear.[178] The key question was whether nonlinearities at the molecular scale were important to the overall pattern of deposition. In 1983, the NRC published an assessment proposing that the answer was no:

> On the basis of currently available empirical data and within the limits of uncertainty associated with the data and with estimating emissions, we therefore conclude that there is no evidence for a strong nonlinearity in the relationships between long-term average emissions and depositions in North America.[179]

This finding was based on certain key assumptions, including that the amount of other chemical species in the atmosphere remained unchanged and that long-term average emissions and deposition were the main concern. These assumptions were considered by other assessments to be too restrictive to permit a generalized statement in favor of linearity. The differing perspectives derived from whether the assessments concentrated on a long-term ecosystems perspective or a short-term chemistry perspective and whether or not they viewed the uncertainty related to chemical reactions in the aqueous phase as decisive (see table 2.3).

NAPAP's assessment of whether source-receptor relationships were linear changed over the course of the program. The annual report concluded in 1983—with some restrictions—that there was likely "a near one to one (linear) relationship between sulfur dioxide emissions and sulfate deposition" in eastern North America.[180] This statement was probably influenced by the NRC report that came to the same conclusion earlier that year. Both assessments referred to averages on broader time scales and geographic areas, thus taking the perspective of ecosystem scientists.

EPA senior scientist Robin Dennis, former leader of the Task Group on Atmospheric Transport, noted that NAPAP scientists had largely settled the question of nonlinearity by 1985. It was shown for the gas-phase oxidation of SO_2 that there was no strong, but only a slight, nonlinearity.[181] However, this only held for gas-phase chemistry and not for aqueous-phase reactions. Chapter 4 of the 1987 Interim Assessment suggested that the scientific evidence indicated the importance of nonlinearity: "Complex interdependencies among chemical formation, chemical reaction, and removal rate processes of pollutants" as well as "variable weather patterns . . . are likely to impose nonlinearity into the SRR's [source-receptor relationships] for sulfur and nitrogen pollutants." It further announced that NAPAP would pursue more studies on the role of nonlinearity and concentrate its efforts on cutting-edge modeling.[182]

The Integrated Assessment also emphasized nonlinearity, as the "transport, chemical, and scavenging processes in the atmosphere are generally nonlinear. This means that the relationship between changes in emissions and changes in exposure can vary for different emission levels."[183] In the SOS/T report 2, however, much more space was given to a discussion of the circumstances under which linearity and nonlinearity may be important, and the report concluded with an emphasis on the importance of linearity when larger time spans and regions are considered.[184] Why did NAPAP's assessment of linearity change?

Disciplinary Differences

In interviews, several NAPAP authors stated that the different disciplinary approaches to the issue strongly affected the debate over linearity. While chemists were focusing on nonlinear chemical reactions that occurred within short time spans of seconds, hours, and days, ecologists were focusing on larger seasonal averages and effects of acid deposition. Malanchuk explained the viewpoint of a systems ecologist on the linearity question:

> Sometimes the more nonlinear they [ecological systems] are, the more linearly they behave, so couldn't we just forget all this stuff and use a linear representation

TABLE 2.3. Relationship between emissions and acid deposition (emphasis added).

Source	Quotation
Canada/United States Coordinating Committee, *United States–Canada Memorandum of Intent on Transboundary Air Pollution: Final Report*, vol. 2	"Over the *shorter time and space scales*, all of the important sulfur dioxide chemical conversion processes are *non-linear.*" (11.15)
NRC, Committee on Atmospheric Transport and Chemical Transformation in Acid Precipitation, *The Acid Deposition: Atmospheric Processes in Eastern North America*	"There is *no evidence for a strong non-linearity* in the relationships between *long-term average* emissions and deposition." (7)
US EPA, *Acid Deposition Phenomenon and Its Effects: Critical Assessment Review Papers*, 2, vol. 1	"In a fresh plume with high concentrations of SO_2, OH level is significantly controlled by SO_2 itself, and the oxidation of SO_2 *is a nonlinear process*. . . . If there are no further fresh injections of SO_2 into this plume, the formation of OH will be governed by the NO_x-HC chemistry in the plume and by entrainment from the background of OH itself and of other reactive species contributing to OH formation. The direct dependence of plume NO_x-HC chemistry on local SO_2 concentration is very weak in this stage of plume transport. Consequently, one commonly finds in the published literature explicit or implicit statement[s] about *linear sulfur chemistry* under such conditions." (4-72)
NAPAP, *Annual Report to the President and Congress (1983)*	"Based on the limited information available, it appears that when *averaged over the entire eastern half of North America* for a year or more, *a nearly one-to-one (linear) relationship exists* between sulfur dioxide emissions and wet deposition of sulfate. Over smaller time and space scales, though, this relationship may not hold." (2)
RMCC, *Assessment of the State of Knowledge on the Long-Range Transport of Air Pollutants and Acid Deposition: Part 1: Executive Summary*	"Over the past few years, simple linear transport models have been refined to the point that wet deposition of sulphur over a one-year period is reasonably well simulated, indicating that *the non-linear character* of sulphur chemistry in the atmosphere *is not critical* in this application." (1-8)
NAPAP, *NAPAP Interim Assessment*	"A reduction in the emissions of sulfur dioxide in the northeastern quadrant of the United States in winter *is unlikely to result in a proportional decrease* in the formation and subsequent deposition of sulfuric acid over the northeastern United States. The magnitude and geographical extent of this so-called 'nonlinearity' have *not yet been evaluated*. Thus, although reducing the emissions of sulfur dioxide in any season is likely to result in the reduction of dry and wet deposition of sulfur compounds, the magnitude and extent of the reduction are uncertain." (1:I-8)
NAPAP, *1990 Integrated Assessment Report*	"Current emissions and deposition data show a relationship between the region of large SO_2 emissions and regional sulfate wet deposition." (167) "Model results indicate that *when the total annual SO_2 emissions are reduced by 50%, total annual sulfur deposition* attributed to those emissions *are reduced by nearly the same amount*." (168)

> of all these nonlinear processes? And I think you probably could have. But (the chemists) would say: no, it does not accurately reflect the science.[185]

Bruce Hicks, a meteorologist who was head of the Atmospheric Processes Task Group and lead author for the SOS/T report 2 that dealt with the question of source-receptor relationships, pointed to another important disciplinary difference: the difference between a focus on effects and a focus on chemical reactions: "If you look at receptors, it [nonlinearity] doesn't matter that much: differences, yes, but within the measurement error. Scientific disagreement will likely persist."[186]

The scientific disagreement was not about whether nonlinearity existed but about whether it mattered for the assessment. Scientists from different disciplines accepted that the chemical reactions were often nonlinear. However, whether or not nonlinear reactions needed to be taken into account in order to make statements about deposition patterns and acid deposition effects became a disputed question among NAPAP scientists. Hales noted in an interview: "The linearity versus nonlinearity was sometimes a passionate debate. We didn't really understand enough of the problem to really tell if nonlinearity was an important aspect or not."[187]

The question of how important nonlinear reactions were for assessing transport patterns and effects of acid deposition was a scientific question that was extremely close to the political question of how much certainty was required to take action and how detailed knowledge had to be in order to provide a basis for regulation. And yet, as much as the scientific debate was shaped by the different disciplinary approaches to the issue, the scientists' views on linearity cannot be reduced to whatever political views about acid rain regulation they may have held.

The Development of RADM

Two different motivations supported the development of RADM: atmospheric chemists and modelers wanted to develop a model that could include the nonlinear chemical processes that they considered to be important for an evaluation of regional deposition patterns, and agency officials wanted a model they could use to respond to policy makers' assessment needs.

The main objective of RADM was the development of "a model that incorporates all known major physical and chemical atmospheric processes related to acidic deposition to provide a scientific basis for estimating the change in deposition due to major changes in precursor emissions."[188] The model represented mathematically the nonlinear processes of oxidant formation (discussed above) from precursor emissions of NO_x and hydrocarbons, gas-phase conversion of SO_2 to sulfuric acid by reaction with oxidants, atmospheric transport of all of these pollutants over long distances, and the nonlinear processes in the response of wet deposition of SO_2 to a change in emissions.[189]

RADM seemed to be everything that US policy makers would need in order to make decisions on acid rain regulation: its objectives were to calculate the changes in deposition over the eastern United States and southeastern Canada that would result from changes in emissions over this region in the next 25 to 50 years, to use projected changes in seasonal and annual total deposition to assess the effectiveness of different emissions control options, and to provide input into models that would calculate the effects of different levels of acid deposition on ecosystems. In addition, RADM could be used to give estimates of how much the emissions in one region contributed to acid deposition in another.

In November 1989, the GAO emphasized RADM's relevance for making political decisions about a 10- or 12-million-ton reduction of SO_2 emissions being debated in Congress at that time "without scientifically sound knowledge of the extent to which, and where, actual deposition would decrease as a result of such controls."[190] These high expectations for RADM's policy relevance did not come to fruition. In the end, neither RADM results nor NAPAP scenarios had any substantial influence on policy makers' decisions about how many tons of sulfur needed to be removed from the atmosphere.

Linkage of the Models

The ambitious linkage of RADM and models that simulated effects of acid deposition on, for instance, aquatic and terrestrial ecosystems did not happen. In its comprehensive report on the use of models for the final assessment, NAPAP reported "some important spatial and temporal mismatches" between RADM outputs and the inputs needed for the effects models.[191] There were attempts to overcome these mismatches, but these were only partially successful, as many of the scientists in the task groups focused on their respective areas of expertise and were not primarily interested in facilitating the assessment. There was no "push" from the side of the researchers involved in doing the assessment in making the data compatible, as Malanchuk noted:

> It was difficult to get this done because we had a lot of people who . . . were interested in emissions when they were in the Emissions Task Group but they

> were less [concerned about] what went on in the Aquatic Task Group, and we go into some of these meetings and we say, "Do you realize that some of the outputs of the emissions model isn't the same as the input to the atmospheric model? Doesn't that bother you somehow?" And they would say, "Hell no, doesn't bother us."[192]

Another reason the linkage of models proved to be highly difficult was model uncertainty. As Hales explained, "We never really did get a good handle on the uncertainties of our models and that was probably one of the key weaknesses."[193] In particular, the coupling of models would have increased the uncertainty tremendously, as Malanchuk pointed out:

> If you would hook the various models together and look at the uncertainty that was generated, you would have no confidence in anything. . . . We had a person on the NAPAP staff whose purpose it was to sort of do the uncertainty analysis of the integrated model. Never happened. Because the uncertainties were just so huge.[194]

NAPAP's Scenarios and the Question of Facts and Values

The scenarios that would be produced with the linked models were to be NAPAP's key contribution to policy making. The scenarios were also the area where the question of the difference between providing policy-relevant knowledge and making policy recommendations became most visible. For the development of the scenarios, many value-laden decisions had to be made, ranging from decisions about which processes should be included in the models to decisions about what energy supply and consumption would look like in the next decades. How did the scientists handle the question of producing knowledge that is relevant for policy makers without suggesting specific policies, and what did they actually do when they produced the scenarios?

A common view on the role of science in policy making is that "if science is to be trusted and have an impact on the national scene, scientists must make a concerted effort to separate fact from policy judgment."[195] Yet this apparently clear separation can easily become blurred. Where precisely should the line between "fact" and "policy judgment" be drawn? The choices that scientists had to make in order to develop the models, and in particular in order to develop the scenarios for NAPAP's final assessment, provide an example of how these two objectives can become intertwined.

In his account of how the scenarios for the Integrated Assessment were developed, Malanchuk relates that the scientists were well aware of the policy implications of the choices they had to make about what to include in or exclude from the scenarios:

> They were doing the prescriptive stuff in their head even if they weren't talking about it because they were looking at it and saying, "Oh boy, the implications of that scenario aren't really very good." . . . At that point in time things that had a huge influence was what emissions scenarios were you going to run. What was the world going to be like in the future? . . . Were people being genuine or were they angling for their preferred scenario because they knew it would produce their preferred result? And did the scientists get manipulated and somehow used in this scenario to help one side or the other by using their science to get them to say one thing or another?[196]

One particularly controversial question was how emissions would change in the future. The answers depended heavily on assumptions about future energy scenarios (for instance what the lifetimes of existing high-emitting plants are, to what extent clean coal technologies would be used in the future, and how high the percentage of non-fossil fuels in the energy mix would be), but also about demographic developments and changes in energy demand. In its 1989 review of RADM, the GAO mentioned a disagreement between EPA and DOE over the extent to which clean coal technologies would be adopted by the utility industry in the future.[197]

In hindsight, as EPA scientist Robin Dennis noted, NAPAP's base-case scenario against which the different control scenarios were evaluated was "tremendously optimistic."[198] The scenario assumed that in the absence of legislation, SO_2 emissions would increase moderately until 2005 and then decrease rapidly to the year 2030. It assumed the gross national product would grow 2.56% per year until 2010, followed by more moderate growth of 1.65% per year until 2030. The lower growth rate for gross national product led to the assumption of lower energy use and emissions. The base-case scenario made similarly optimistic assumptions about the future energy mix, the lifetimes of power plants, the retirement rates of plants, and the adaption rates of low-emission technologies. For example, the scenario assumed that by 2010, highly efficient integrated gasification combined cycle (IGCC) plants would produce 73,000 MW of energy.[199] In 2012, however, there were two operating IGCC plants in the United States, supplying 250 and 262 MW of energy to the electricity grid, respectively.[200] The costs of the CAAA emissions reductions, however, have been lower than those estimated by NAPAP. These reductions are now projected to be $1–$2 billion dollars per year,[201] while NAPAP had calculated \$2.5–\$3.5 billion per year for this scenario.[202]

Challenge 3: Time Pressure—Who Is Doing the Assessment?

Throughout the assessment, it was often unclear who was in charge or even what an assessment was. While it may seem odd that questions as fundamental as these were not well defined, this disconnect can be attributed to the hybrid construction of NAPAP as a research program and an evaluation of impacts and policy options. While there was consistency in the former, there was no consistency in the handling of the latter. In the beginning, the Assessment Task Group was in charge of the assessment, but there was little motivation to do it, in part because research results had yet to be produced. When Kulp took charge of NAPAP, he scaled evaluation back drastically. Finally, a core group at NAPAP's headquarters in Washington were primarily concerned with getting the pieces for the Integrated Assessment together rather than with writing the report.

So who was doing the assessment in the end? The authors of the Integrated Assessment were the same scientists who were writing the SOS/T reports.[203] The idea behind having the same people do the survey on the science and then the policy-relevant evaluation of this science was to guarantee the scientific integrity of the final document. There were 67 contributors to the Integrated Assessment, all of whom were authors of the SOS/T reports. That the scientists in NAPAP produced two different documents largely at the same time led to confusion about what NAPAP's "assessment" actually was. Bruce Hicks, a task group leader, stated: "A big topic was 'What is an assessment?' At the task group level, people thought it is an assessment of the science. The policy assessment was above the task group level . . . done by the people in Washington."[204] The "people in Washington," however, had decided that "the task groups would have the responsibility to produce their chapter in the Integrated Assessment" (see figure 2.5).[205]

In 1990, the question of who should be responsible for doing the Integrated Assessment was a major issue of concern. NAPAP's Oversight Review Board (ORB) evaluated NAPAP before the Integrated Assessment was published. Milton Russell and Kenneth J. Arrow, chairman and member of the ORB, respectively, were concerned about whether NAPAP could meet its timeline for producing not only the SOS/T volumes but also the Integrated Assessment. Arrow wrote to Russell on March 2, 1990:

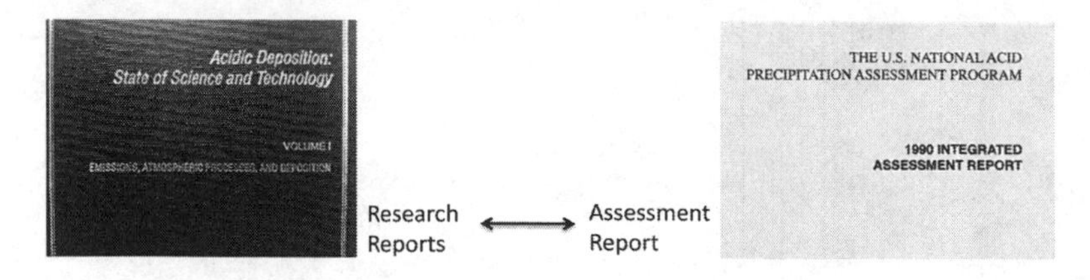

Acid Precipitation Act

"(1) to identify the causes and effects of acid precipitation"

"(2) to identify actions to limit or ameliorate the harmful effects of acid precipitation"

Conflicting views on what the assessment is

When scientists thought they were doing an "assessment of the science," they were referring to the State of Science reports.

When NAPAP's headquarters wanted the scientists to produce an "assessment for policymakers," they were referring to the Integrated Assessment.

FIGURE 2.5 Conflicting views on NAPAP's goals.

> You spoke of the time pressure on the researchers who are both going to complete the SOS/T documents *and* write the assessments and summaries. Is it too late to bring in other personnel to help with the latter two? This might have the additional advantage of fresh perspectives which would introduce the policy-oriented wisdom you have found so lacking.[206]

Russell was skeptical about this arrangement. He reported to the ORB from the final international NAPAP conference that took place at the Hyatt Regency Hilton in South Carolina on February 12, 1989:

> This Friday morning integrated findings session was a great disappointment to me. With few exceptions, the presenters simply talked about relatively narrow scientific issues on which there was either agreement or disagreement. They did not relate to the broader themes that are relevant to public policy. There was no indication that these persons are thinking in terms of the "wisdom" that must come out in the Integrated Assessment, nor were they thinking of the hierarchy of questions that would be addressed in the Integrated Assessment itself. Unfortunately, *these are the persons* who have the responsibility of drawing the Integrated Assessments together.[207]

Russell's observation indicated that even at the end of the project, many researchers were focusing much more on the SOS/T research reports than on the Integrated Assessment.

Challenge 4: "Push and Pull"—Who Wants the Assessment?

According to Malanchuk, scientists' focus on the research reports led to delays in handing over research results, which in turn led to a significantly shortened time frame in which to complete the Integrated Assessment:

> You have to have the work done at such or such a date in order to do the Integrated Assessment, and you just run out of time. . . . All the scientists are saying, "My future is on the state of the science reports, not on the assessment, and I don't care anything about the assessment, I don't care about policy-relevance, I only care about doing good science." So they, these several hundred scientists, were mostly interested in the SOS report; they weren't at all interested in doing an Integrated Assessment. And when it came time to do the Integrated Assessment, there wasn't enough time; the scientists weren't turning it over to the Integrated Assessment team unless they were involved, and so everything got pushed, so doing a real Integrated Assessment at the end never happened.[208]

The high-level officials in the agencies were not pushing for the Integrated Assessment, although for a different reason than the agency scientists. Mahoney experienced this lack of interagency support as the main obstacle to getting the Integrated Assessment thoroughly done:

> In my own view on the years that I worked with the program there was one major shift in the views that I perceived, which was that everybody was very supportive at the beginning, that we do the right kind of things, we get the science back on track, so I had no dispute about that, but after a year, maybe 18 months when it was time to start really looking at alternate scenarios . . . more agency representatives started expressing great concern, almost fear, about [doing] anything like that. It was like: "Go play your science game, but don't come in here with something that sounds like you are giving us answers." . . . I wouldn't say that this was universal, but I was starting to get a lot of cautionary feedback from these high-level people in the agencies, not the working scientists so much. . . . To my mind, the biggest limitation and disappointment for me in what we were able to get done was the limitation about really moving forward with doing comparative analysis, because I didn't have the support on an interagency basis to do those cases. . . . There was a lot of ground-breaking research . . . a lot of good work, but the idea of putting it all together was scaring a lot of people.[209]

This lack of agency support for the Integrated Assessment, present at NAPAP's beginning, severely hampered the success of NAPAP in the end.

It is not entirely clear why the agencies did not support the comparative analysis of different scenarios. In the beginning of NAPAP, they did not support policy-relevant analyses because the Integrated Assessment was years in the future and the program operated in a regulation-hostile political environment. This was no longer the case when the report came due. The reason for this lack of support at the end of the program could be that at the time when NAPAP started to compare the effects of different emissions reduction scenarios, the 10-million-ton SO_2 emissions reduction that was finally envisioned in the CAAA was more or less already decided. If NAPAP had published results that indicated that other amounts of emissions reductions, the inclusion of other pollutants, or a reduction on a different time scale than indicated in the CAAA would lead to better outcomes (both in terms of costs and in terms of environmental benefits), this would have left policy makers with a big problem.

The Impact of NAPAP's Final Assessment

There is considerable disagreement whether NAPAP as a whole was a success. This disagreement partly reflects whether or not people were involved with NAPAP but also results from different understandings of what counts as a success. In what follows, we do not attempt to measure NAPAP's success. Rather, we discuss what different actors understand by success and in which different ways assessments can be linked to policy.

We found general agreement among the NAPAP participants and also in the secondary literature that NAPAP was a success in terms of its research program. NAPAP advanced the scientific understanding of acid rain; examples of important NAPAP research results emphasized by the interviewees and in the literature include the direct-delayed response hypothesis in the area of aquatic effects research,[210] the RADM, the NSWS, and the Paleoecological Investigation of Lake Acidification.[211] However, we also found agreement among interviewees and the literature that NAPAP did not fundamentally change the basic scientific understanding of acid deposition. Of course, it was not within NAPAP's original task to "greatly [change] the world view of acidic precipitation," as Schindler implies.[212] Yet the question of whether NAPAP substantially advanced knowledge about acid deposition included, often implicitly, another question—namely, whether or not the development of acid rain regulation required the knowledge NAPAP accumulated.[213]

NAPAP was not a research program but an assessment based on a research program. This leads us to ask: Were the scientific advances fostered by

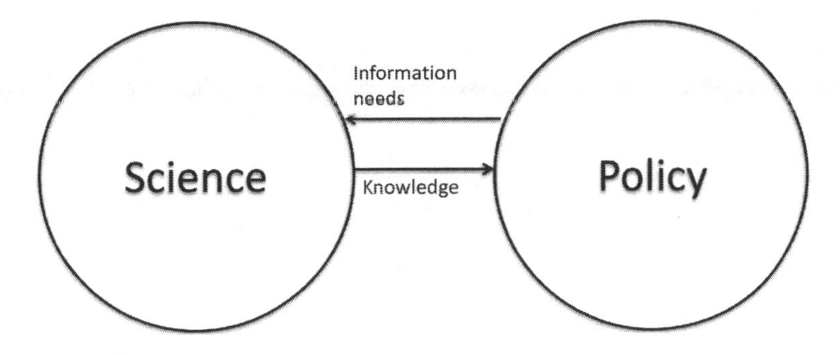

FIGURE 2.6 The traditional view of the science-policy linkage.

NAPAP related to NAPAP's policy relevance and, if so, in what ways? When it came to the question of whether NAPAP was a successful assessment, the views of the interviewees were more negative, with one important difference: NAPAP scientists (in interviews and in their published accounts of NAPAP) had more positive and differentiated views on NAPAP's policy relevance than did the general literature, which concluded in large part that, as an assessment, NAPAP was a failure.[214] We argue that this difference in views resulted mainly from differing understandings of what counts as success, while also taking into account the often positive bias of scientists (or anyone else) toward an assessment (or any other activity) in which they took part.

When NAPAP published its final assessment in 1991, the CAAA had already been passed (November 1990). Does this mean NAPAP was not successful? This negative view was expressed by Edward Rubin, Ralph Perhac, and the ORB of NAPAP.[215] However, this sentiment reflects only one specific definition of success: the clear and visible impact of an assessment's end results on legislation. According to this view, science and policy are clearly separated areas. The linkage of the two arenas works through the reciprocal communication of scientific results and political information needs. Consequently, proponents of this view see science-policy linkages primarily as communication problems (see figure 2.6).

This traditional view of the assessment-policy linkage is restrictive, as it does not take other forms of influence into account. In particular, this view neglects any formal or informal communication that may take place before the release of a final report. A more complex view, however, acknowledges

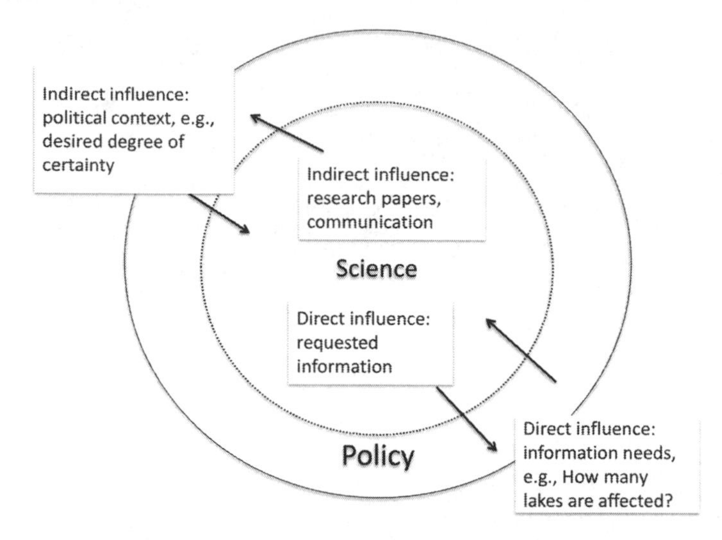

FIGURE 2.7 The semipermeable science-policy linkage.

that science and policy are not clearly separated. Rather, the two have a semipermeable boundary through which knowledge and information needs, in ways both formal and informal, may "seep" (see figure 2.7). In this way, science is always influenced to some degree by its political context, and the political context likewise is affected by the knowledge that seeps through various channels into political debates.

Not only were NAPAP results widely distributed before the publication of the Integrated Assessment but NAPAP director Mahoney had provided input during the negotiations for the CAAA:

WAZECK: Some people argue that NAPAP failed because the Final Assessment appeared after the CAAA passed. How do you see the relation between NAPAP's final assessment and the legislation?

MAHONEY: Doesn't make any difference. Actually, there was plenty of communication; you don't need the written word in the report. . . . I attended many of the informal, but very pointed, domestic policy council discussions in the shaping of this bill [the CAAA]. . . . I was very well heard in all of these things.[216]

Similarly, several other interviewees pointed out that NAPAP's influence on policy making was to be found in its piling up of evidence and not in any major breakthrough in acid rain research. For example, Malanchuk said:

> I can't say that there was one of these eureka moments, because of the science the policy result became crystal clear; that didn't happen, but I would say that there was a weight of evidence that basically said, . . . "OK, we know that this sulfur dioxide wasn't in the air to begin with, we really ought to get it out." We could have done that in 1983 or 1982 and the 10 or 7 or 8 years of research more or less provided the weight of evidence for us to go along with this scenario and we would say, "OK, is it 5 million, or 10 million, or 15 million, and at the end of the day we'd say, well, the most we can possibly get any political support for to get the CAAA passed would be 10 million tons. You can't get more than that." . . . Was that wrong? No. I don't feel bad about 10. And I felt that NAPAP helped support the decision about 10.[217]

When the CAAA passed, many saw it as a major accomplishment in addressing the problem of acid rain.[218] The CAAA provided the basis for a 10-million-ton reduction of SO_2 emissions and introduced a cap-and-trade system for emissions. Mahoney tried to convince policy makers to opt for a 12-million-ton reduction by 2002, but "everybody can understand 10 million tons by 2000, so that became the only policy prescription that came out of the acid rain things at the end of the day."[219] Malanchuk likewise supported Mahoney's statement that the 10-million-ton reduction was a political decision, pointing out that it was based on political rather than scientific considerations: "Why 10? It wasn't because of any scientific analysis. I have heard people say, 'We have to have a two-digit number to make the Canadians happy.'"[220] However, it is worth noting that even though the selection of 10 million tons was due more to political than to scientific considerations, the number was nonetheless scientifically defensible given the state of knowledge of the time.

RADM, which had been developed to provide policy-relevant knowledge, had no practical influence on the formation of policy. RADM results were not instrumental in the decision about how many tons of sulfur dioxide should be taken out of the air. Jay Messer noted that the 50% reduction target for sulfur emissions that NAPAP supported "was exactly what was envisioned in a 1981 National Academy of Sciences report."[221] The eventual irrelevance of RADM's results became a major point of criticism; as David Hawkins said: "It might be a feat of computation, but who cares, from a policy standpoint?"[222]

Of course, policy makers did care in the years before. Mahoney pointed out in 1991 that "the extraordinary effort that went into RADM was justified because the model was designed to answer a vexing—and critical—question:

whether a reduction in sulfur dioxide emissions would lead to a similar reduction in acid rain."[223] But when the administration changed and political priorities shifted, policy makers made a political decision. Herrick and Jamieson also emphasized the importance of changing political priorities, pointing out that the conventional NAPAP critique "presumes that policy makers will define the problem in the same terms; that their 'needs' are transparent and remain constant as time, circumstance, and political fortune change."[224] In the case of acid rain regulation, the political dynamics—namely that a new president, George H. W. Bush, and Senator George J. Mitchell pushed the legislation—had a more significant influence than did the reduction of scientific uncertainty. The new administration acted on the basis of knowledge about acid rain that was not significantly different from what was available a few years earlier. What had changed was the policy makers' view on the sufficiency for action of this state of knowledge.

A more indirect way in which an assessment can be successful is by having a positive influence on future assessments. This can occur when tools developed in one assessment can be applied in other assessments. In this vein, Malanchuk suggested that NAPAP may have helped to establish natural resource assessment:

> This may be a little too ambitious, but I would say that NAPAP helped advance economic modeling for environmental contamination. A lot of things got funded to do the economic analysis of the costs and benefits of acid rain that have helped advance that whole notion of natural resource damage.[225]

This influence can also occur when scientists from one large-scale assessment later become involved in another assessment, bringing with them their prior experience. Mahoney, for example, was from 2002 to 2006 the first director of the Climate Change Science Program. His NAPAP experience had a "very big" impact on how he steered that program's assessments (see table 2.4).[226]

CONCLUSION

From NAPAP's beginning, the program suffered from a structural tension between its research and policy dimensions. Different actors—agency heads, agency scientists, and policy makers in the late Carter and early Reagan administrations—all put a strong emphasis on research and showed less concern with policy.

TABLE 2.4. Different understandings of a successful assessment.

Focus	*Direct Success*	*Indirect Success*
Assessment	Assessment fulfills its task as stated in its program	Assessment influences other assessments Participants gain experience in how to do assessments
Policy	Assessment directly influences legislation	Assessment results diffuse through different channels during the course of the assessment Assessment serves as a reference point for the policy discourse
Science	Assessment advances the state of knowledge	Assessment sets new reference points for research Assessment fosters communication between scientists

On the basis of different motivations—researching the topic, providing a solid knowledge basis for policy, and delaying regulation—policy makers and others involved in government in the United States in the late 1970s and 1980s perceived acid rain as an important research topic but (with some important exceptions) not as a priority for immediate action. This was particularly true for the agencies and agency scientists who conducted NAPAP. The actors involved in NAPAP exerted no "push" to focus on the assessment aspect. In addition, NAPAP's changing political context accounted for the weak focus. While created by Congress under the Carter administration, NAPAP carried out most of its research and assessment activities under the Reagan administration, where acid rain regulation was not a priority. Consequently, neither the administration nor high-level agency officials exerted any significant political "pull" for NAPAP to focus on the assessment dimension.[227] The "pull" only emerged after NAPAP released its Interim Assessment, stirring a controversy that involved the Canadian government and, consequently, the White House. Then the administration showed a keen interest in getting the NAPAP controversy out of the news and ensuring that NAPAP would publish an uncontroversial and scientifically sound final report.

In order to get the Integrated Assessment done, the scientists at NAPAP headquarters had to deal with substantial problems: first, they had to establish assessment procedures and an assessment framework; second, they had to make enormous efforts to integrate into a final report the various pieces of research that NAPAP projects had produced; third, they had to produce the assessment report under high time pressure; and fourth, they had to do so with insufficient support from the agency scientists, the agencies themselves, and the administration.

Although the group around Mahoney and Malanchuk at NAPAP headquarters spent considerable time and effort attempting to combine the research results from 1988 onward, the final report did not integrate the compartmentalized research. NAPAP had established the assessment framework too late, and the strong research orientation within the program still prevailed. The integration of data provided by the different major NAPAP research projects and the linkage of models crucial for producing the different scenarios included in the final report proved to be particularly difficult.

In order to investigate the consequences of this research orientation for the final assessment, this chapter put special emphasis on the study and modeling of the relationship between sources of emissions and acid deposition at specific sites—a scientifically challenging issue that was a core area of NAPAP's activities and that was at the center of the US policy debate on acid rain in the 1980s. The Regional Acid Deposition Model was one of the major NAPAP-funded research projects, and it was supposed to elucidate source-receptor relationships. RADM became both the center and the bottleneck of the final Integrated Assessment: it was unclear whether the model would be usable on time, whether the data necessary to run the model could be provided, or whether RADM could be linked to models of acid deposition effects. In hindsight, RADM, which was supposed to deliver the answer to the most relevant policy question of acid rain regulation in the 1980s, turned out to be irrelevant for the 1990 CAAA. Yet it did substantially advance atmospheric modeling; today, European and Canadian scientists also use RADM-type models.

The relevant aspect of NAPAP's Integrated Assessment was not that it was published too late—there was considerable informal communication when the CAAA regulations were designed—but that it produced answers to questions that turned out not to be decisive in the design of regulations. Policy making on acid rain was substantially influenced not by the new evidence NAPAP provided but by changes in political priorities. This raises the question of what precisely is the role of scientific assessments in the formation of environmental policy. The history of NAPAP suggests that this role is complex, difficult to measure, and potentially indirect.

Compared to other US and Canadian acid rain assessments, NAPAP had a stronger inclination to assess the extent and effects of acid rain in the United States conservatively. It concluded more often that specific issues were uncertain, and framed the evidence in ways that suggested that acid rain was not a major environmental problem. This particularly strong inclination to "err on the side of least drama" can be traced back to the influence of the agencies, particularly the DOE, on NAPAP reports, and to the awareness of scientists who took part in the program of the likelihood that their conclusions would be drawn into political debates. Although we have not studied what influences affected Canadian or European scientists in their respective assessments, these could have been present and operating in a different direction.

Much of our analysis has dealt with NAPAP's peculiarities. However, an understanding of the factors that shaped NAPAP's development may provide more general insight into the factors that influence the performance of environmental assessments. Specifically, our analysis suggests that attention should be given to the interests of the actors in doing the assessment (push), the interests of policy makers or the public in receiving the assessment (pull), and the changing information needs and political priorities that can change both the external relevance of an assessment and its internal dynamics. The history of NAPAP shows how the political context of an assessment may influence how the involved scientists define research priorities, as well as the ways in which they evaluate scientific uncertainty.

APPENDIX 2.1. KEY ISSUES COMPARED ACROSS ASSESSMENTS

1. Historical trends in acid deposition: Does there appear to be a historical increase in the trend of quantity or intensity of acid deposition in North America?
2. Historical trends in SO_2 emissions: Does there appear to be a historical increase in the trend of quantity or intensity of SO_2 emissions in North America?
3. Future trends: Are increases in acid deposition or its precursors predicted for the future?
4. Sources: Do anthropogenic sources account for a major amount of emissions of acid rain precursors?
5. Dry deposition: Is the damage resulting from dry acid deposition comparable to that of wet?
6. Aquatic effects: Is there evidence that acid deposition substantially damages aquatic ecosystems?

7. Terrestrial effects: Is there evidence that acid deposition substantially damages terrestrial ecosystems?
8. Health effects: Is there evidence that acid deposition substantially affects health?
9. Materials effects: Is there evidence that acid deposition substantially damages materials?
10. Long-range transport: Can the precursors of acid deposition, or the resulting acids, be transported long distances from their point of origin?
11. Point sources: Can damages to specific areas be attributed to point sources of emissions?
12. Source-receptor relationships: Is it possible to establish relationships between regions with high emissions and regions receiving acid deposition?
13. Linearity: Is the relationship between emission sources and acid deposition linear?
14. Models: Are transformation and transportation models precise enough to serve as the basis for the development of control strategies?
15. Political action: Are actions to reduce acid deposition required? Should governments act now?

ACID RAIN TIMELINE

1950s–1970s

- Acid rain gradually defined as a problem, first regionally, then internationally (especially with growth of tall smokestacks toward the end of this period)
- US Clean Air Act passed in 1963, 1970

1972

- UN Conference on the Human Environment in Stockholm, Sweden
- Norway's SNSF acid rain research program begins; runs to 1980

1977

- OECD *Programme on Long Range Transport of Air Pollutants: Summary Report* published
- Major amendments to the US Clean Air Act

1979

- President Carter establishes the Acid Rain Coordinating Committee to develop a coordinated US federal acid rain program

- UN Economic Commission for Europe Convention on Long-Range Transboundary Air Pollution signed by 34 countries, including the United States and Canada

1980

- Norwegian SNSF final report published
- Memorandum of Intent Concerning Transboundary Air Pollution established between United States and Canada
- NAPAP created by US Congress (first director: Chris Bernabo)

1981

- National Research Council report calling for prompt action to reduce atmospheric emissions

1985

- Helsinki Protocol establishes binding emissions reductions of at least 30% from a 1980 baseline
- Canada and many European nations sign the Helsinki Protocol, but United States does not
- Bernabo steps down; J. Laurence Kulp appointed new director of NAPAP

1987

- Helsinki Protocol comes into force
- *NAPAP Interim Assessment Report* published; Kulp's executive summary criticized for downplaying or even misrepresenting the report's actual findings; Canadian environment minister denounces it as "voodoo science"
- Kulp resigns; James Mahoney appointed new director of NAPAP

1990

- Clean Air Act Amendments passed, providing the basis for a 10-ton reduction of SO_2 emissions and introducing a cap-and-trade system for emissions

1991

- *NAPAP Integrated Assessment Report* published

CHAPTER THREE

Assessing Ozone Depletion

INTRODUCTION

The Montreal Protocol to limit ozone-depleting substances (ODSs) in the stratosphere is often touted as a model of successful science-based policy.[1] ODSs are versatile chemicals involved in a range of industrial and commercial applications and processes, and the international agreement to limit their production and release antedated the full development of commercially viable alternatives. Environmental law and policy scholar Edward Parson has argued that "the ozone issue highlighted the crucial policy influence of official scientific assessments, as distinct from scientific results themselves."[2] Former ambassador Richard Benedick also concluded that scientific assessments were crucial in influencing ozone policy.[3] To paraphrase Benedick, scientific theories and discoveries alone were not sufficient to influence policy, but scientific theories and discoveries as articulated and expressed in assessments went hand in hand with, and played an essential role in, informing and building policy.

Parson and Benedick have both written eloquently about the influence of assessments on ozone policy, but what do we know of the assessments themselves? What did scientists actually do to assess the science? How did they understand what they thought they were doing? By reviewing the work done by scientists involved in the ozone assessment process, this chapter attempts to answer these questions.

We begin by tracing the development of ozone science and the emergence of scientific concern over the possibility of ozone depletion. We then track the evolution of scientific assessments of ozone, beginning with the national assessments of the United States and United Kingdom and continuing with the international assessments that provided the groundwork for the adoption of the Montreal Protocol. We then discuss the ongoing assessment process that has influenced the continuing updating of the Montreal Protocol and highlight some of the general themes that emerge from this investigation.

As the story unfolds, we see the importance of what gets put in and what gets left out of assessments, particularly when substantive aspects of the science are still in flux, and how this matters when policy decisions are at stake. We also see the importance of who participates in the assessment and how decisions are made regarding the appropriate experts to author an authoritative report. We also note how ozone scientists from various disciplines came to favor a consensus approach to assessment. This influential choice has contributed to forging a close relationship between ozone assessment and policy but has also imposed some important limitations.

HOW OZONE BECAME A TOPIC FOR SCIENTIFIC ASSESSMENT

Ozone Science

Ozone is a gaseous molecule composed of three oxygen atoms bound together (O_3). It occurs naturally in the stratosphere—the middle layer of the earth's atmosphere—at an altitude of approximately 15 to 50 km. Ozone is highly reactive, and it is constantly destroyed and created in the stratosphere through a series of chemical reactions driven by ultraviolet (UV) radiation from the sun and the presence of other molecules with which ozone, elemental (O) and molecular (O_2) oxygen can react.

The nature of these reactions explains why ozone is found where it is: sufficient UV radiation to break down oxygen molecules does not penetrate much below 15 km while ozone is destroyed too quickly above 50 km. In fact, most stratospheric ozone is found at altitudes between about 20 and 30 km. This "ozone layer" acts as a natural shield protecting humans from skin cancer, cataracts, and other conditions and protecting all life on earth from the sun's harmful UV-B radiation. A threat to ozone is a threat to most life on earth.

The initial awareness of human potential to damage the ozone layer arose in the 1960s when the United States, France, and the United Kingdom began to consider building a fleet of supersonic transport (SST) airplanes. These aircraft fly at a higher altitude (up to 20 km) than conventional commercial jets, and some scientists and environmentalists expressed concern that water vapor and nitrogen oxides (NO_x) in the exhaust from SSTs might damage the ozone layer.[4] While scientific work suggested that there could be a threat, plans to build the proposed large fleets of SSTs were scrubbed for a variety of reasons before scientific consensus was reached on this question. The SST controversy did, however, raise awareness of potential harms to the ozone layer, and paved the way for later scientific work that took these harms into consideration.

A 1974 paper by US National Aeronautics and Space Administration (NASA) scientists Richard Stolarski and Ralph Cicerone proposed a gas-phase catalytic cycle or chain reaction in which oxides of chlorine (ClO_x), if found in sufficient quantities in the stratosphere, could destroy significant amounts of ozone.[5] The reactions can be summarized as follows:

(1) $Cl + O_3 \rightarrow ClO + O_2$

(2) $ClO + O \rightarrow Cl + O_2$

(3) $O_3 + O \rightarrow 2O_2$

In reaction 1, chlorine atoms (Cl) react with ozone (O_3) to form chlorine monoxide (ClO) and oxygen molecules (O_2). In reaction 2, the chlorine monoxide created in the first reaction combines with free oxygen atoms (O, produced from the breakdown of oxygen molecules high in the stratosphere exposed to UV radiation) to produce chlorine atoms and oxygen molecules. Reaction 3 shows the net effect of the preceding two reactions: catalyzed by the chlorine, ozone and oxygen atoms are converted to molecular oxygen, while the chlorine atom is left free to participate in reaction 1 once more. In this way, chlorine atoms can be recycled over and over to catalyze the destruction of vast quantities of ozone. (The chlorine eventually gets tied up in the form of hydrochloric acid [HCl] and chlorine nitrate [$ClNO_3$], both of which react very slowly in the gas phase and so can be considered chlorine reservoirs. We will meet these reservoirs again later in this chapter.)

While Stolarski and Cicerone mentioned several potential sources of chlorine, including sporadic volcanic eruptions and exhaust from solid fuel rockets (such as NASA's new space shuttle), at that time there were no known large, human made sources of stratospheric chlorine. Stolarski and Cicerone thus concluded their paper by calling for "a careful reevaluation of all possible chlorine sources." Meanwhile, two chemists at the University of California, Irvine, had discovered a major source of stratospheric chlorine. Mario Molina and his postdoctoral advisor F. Sherwood Rowland published a paper in *Nature* the same year in which they showed how it was possible for Cl to reach the stratosphere in quantity via chlorofluorocarbons (CFCs), where it could participate in exactly the sorts of reactions that Stolarski and Cicerone had anticipated.[6]

CFCs had many applications; the best known and most important were as coolants in refrigeration and air conditioning, as aerosol propellants, and as cleaning solvents. When first developed, they were prized for their inertness, nonflammability, and nontoxicity: they do not easily catch fire or react with other substances. This very inertness, however, is also their Achilles' heel, because CFCs released into the atmosphere are not destroyed, altered, or washed out but instead make their way to the stratosphere, where they eventually are broken down by the only thing that does destroy them: the sun's UV radiation. In this way, chlorine is released from CFC molecules into the stratosphere, where—through several chemical reactions including the gas-phase catalytic chain proposed by Stolarski and Cicerone—they destroy the protective ozone layer. When Rowland and Molina's paper was published in the early 1970s, close to one million tons of CFCs were being manufactured worldwide every year.[7]

Stolarski and Cicerone ended their paper on the ozone-depleting chlorine catalytic chain reaction with the suggestion that "the next stage of stratospheric chlorine calculations should include a careful reevaluation of all possible chlorine sources and possible sinks such as adsorption on aerosols [meaning natural/aerosol/particles, not aerosols from spray cans]."[8] Molina and Rowland made a similar point, writing, "Our calculations have been based entirely on reactions in the gas phase, and essentially nothing is known of possible heterogeneous reactions of Cl atoms with particulate matter in the stratosphere."[9]

The scientists were suggesting that one needed to understand both sources of chlorine to the atmosphere and reactions and processes that might remove it from the atmosphere—so-called sinks. One possible sink would be the adsorption and reaction of the chlorine-containing compounds on solid particles or droplets in the stratosphere, such as sulfate particles, which would be "washed out" or "rained out." These types of reactions are called "heterogeneous" because they occur among more than one phase of matter—in this case, gases reacting on solid particles.

The early scientific work and early national assessments on ozone depletion in the mid- to late 1970s took the question of heterogeneous rainout seriously and investigated this possible sink. Ozone depletion might not be a problem if there was major rainout of chlorine compounds but could be a huge problem if there was little or no rainout. A 1976 assessment report produced by the US National Academy of Sciences (NAS) included an appendix on this subject that concluded "that inactive removal of CFMs [a class of CFCs] from the stratosphere by heterogeneous processes is not at all significant," and "heterogeneous removal of ClO is negligible when compared with the homogeneous ozone catalysis processes."[10] In other words, significant rainout of chlorine was not likely, which left the chlorine available to participate in ozone depletion.

Today heterogeneous reactions are viewed as crucial for understanding ozone depletion. Rather than providing a mechanism for protecting the ozone layer, it is now established that heterogeneous reactions involving particulate matter including ice crystals are what makes ozone depletion rapid. Why were heterogeneous reactions dismissed in this early report?

Scientists and assessment authors were right in their judgment that the rainout of chlorine compounds (and other ODSs) on stratospheric aerosol particles was unlikely. There were several reasons to believe this, among them the fact that stratospheric aerosol particles are literally few and far between. The only significant occurrence of particles throughout the global stratosphere is the Junge layer, a thin layer of sulfate particles injected into the stratosphere by volcanic eruptions. This layer of sulfate aerosols occurs at an altitude of about 16 to 30 km. Ozone depletion was expected to occur mainly at an altitude of about 40 km, where UV radiation was sufficient to stimulate key chemical reactions. Scientists therefore concluded that the particles of the Junge layer would play very little role in ozone depletion.

Particles also occur in clouds, but they were also considered unlikely to be significant. Cambridge University chemist Tony Cox explained:

> It was generally held that the heterogeneous reactions in the atmosphere were largely confined to liquid water droplets in clouds where you have got a large body of condensed material, and it was limited to the types of reactions that would operate in cloud water.[11]

Clouds are generally phenomena of the troposphere (the lowest layer of the earth's atmosphere). Even the highest of these clouds are far below the 40 km level at which scientists thought ozone depletion would occur. So if heterogeneous reactions happened in clouds (or on the particles of the Junge layer), and where ozone is depleted there are no clouds (or Junge layer), then heterogeneous reactions would not be significant in ozone depletion.

Moreover, if these heterogeneous reactions did occur, they would have a minor effect, and scientists understandably wanted to focus on what was major. According to Cox, "The atmosphere is complicated enough, so I felt that you didn't really want to complicate it any more by worrying about minor effects."[12] Neil Harris, another Cambridge chemist, agreed, noting that scientists had enough to do trying to figure out the gas-phase chemistry involved with ozone depletion before considering other types of reactions: "It [the gas-phase chemistry] wasn't well known, so you certainly don't go to heterogeneous until you know the gas phase."[13]

This pragmatic approach to ozone chemistry was typical and widespread. It was a form of Occam's razor: keep your analysis simple unless you have evidence to justify making it more complex. Few scientists at the time seem to have objected very much, if at all, to this approach. So they were shocked to discover in the mid-1980s that a different type of heterogeneous reactions did play a critical role. But this realization came only after the effects of these reactions had been detected, in the form of the Antarctic ozone hole.

The Antarctic Ozone Hole

Elaborations of the original Rowland-Molina hypothesis, as well as the gas-phase chlorine catalytic chains proposed by Stolarski and Cicerone and others, suggested that ozone depletion would occur gradually, constantly, globally, and chiefly at an altitude of about 40 km.[14] Field scientists, laboratory workers, and computer modelers did a great deal of work in the 1970s and 1980s to try to predict how much ozone depletion would occur and when it would become detectable. Although predictions of ozone depletion evolved with changing scientific data, theories, and models, by the mid-1980s scientists generally expected to see 5% depletion of the global ozone layer many decades in the future, assuming CFC production and use continued unabated at 1974 levels.

In May 1985, however, the discovery of the Antarctic ozone hole changed the situation dramatically. Announced by members of the British Antarctic survey,[15] the "hole" was a geographically limited circumpolar area of major ozone depletion (up to 60% in total column ozone, i.e., the total amount of atmospheric ozone above any particular area on earth), occurring in the austral spring. Instead of slow, steady, gradual global depletion, scientists had discovered a specific region of radically depleted stratospheric ozone, occurring at one particular time of year. Existing theories provided no explanation for the spatial and temporal focus or for the alarming degree of depletion.

NASA scientists mounted three expeditions to investigate the Antarctic ozone hole: the National Ozone Expedition (NOZE) in August–September 1986, and NOZE II and the Airborne Antarctic Ozone Experiment (AAOE) in August–September 1987. Three main competing theories had been proposed to explain the hole: a dynamic theory in which ozone-poor air was drawn up from the Antarctic troposphere into the stratosphere; a solar theory in which increased solar activity produced large quantities of solar particles that bombarded oxygen and nitrogen, forming ozone-depleting NO_x; and a chemical theory involving reactions of anthropogenic chlorine, nitrogen, and/or bromine.[16] Although the results of these expeditions were not published in peer-reviewed journals until 1989, the initial findings of the AAOE were released on September 30, 1987. These findings supported the chemical theory, strongly suggesting that the ozone hole was caused by perturbed chlorine chemistry involving anthropogenic chlorine, facilitated by the unusual meteorological conditions known as the Antarctic polar vortex, a circular stratospheric flow swirling around Antarctica.[17]

Yet the so-called Blue Books,[18] a major international assessment published in 1986 (almost a year after the discovery was announced) said almost nothing about the ozone hole, because its announcement came while the assessment was being finalized and scientists were still struggling for its explanation.[19] As a sort of placeholder, the following paragraph was included: "The article [i.e., the 1985 paper by Farman, Gardiner, and Shanklin reporting the ozone hole][20] has not yet been assimilated by the modeling community, and it is premature for this report to do more than to note it with great interest and to recommend that it be given close attention in the near future."[21]

Within two years, scientists had determined that anthropogenic chlorine compounds were reacting on the surfaces of icy particles in polar stratospheric clouds, which form each polar winter. These very high-altitude clouds form only at temperatures below −77°C and are variously composed of water, nitric acid (HNO_3), and sulfuric acid (H_2SO_4). Scientists concluded that when sunlight returns to the South Pole in the spring, it sets off a chain reaction in which the chlorine, freed from an inert reservoir by reactions on cloud particles, catalytically destroys most of the ozone in the polar stratosphere.[22] In other words, heterogeneous reactions on polar stratospheric clouds, involving a solid phase—complex ice crystals—play a key role in the chemistry of polar ozone depletion.

Polar Stratospheric Clouds and the Return of Heterogeneous Reactions

Antarctic explorers had observed polar stratospheric clouds, also called noctilucent, nacreous, or mother-of-pearl clouds, for decades; there are records of their observation from Robert F. Scott's British expedition to the South Pole in 1910–1912.[23] The scientists who went to Antarctica in the International Geophysical Year (1957–1958) also observed these clouds. One of them was Joseph Farman of the British Antarctic Survey, who, almost thirty years later, would announce his team's discovery of the Antarctic ozone hole.

Farman was a geophysicist, and chemical reactions, heterogeneous or otherwise, were not his area of expertise. Nor was it the expertise of other scientists who knew about polar stratospheric clouds. Likewise, chemists studying ozone depletion did not know about polar stratospheric clouds. Indeed, the only source of stratospheric aerosols they knew of was the Junge layer—and some, like Rowland, had not even heard of that.[24]

Scientists tended to be segregated by disciplinary specialization. Chemist Neil Harris noted that "there was a division of communities, so the atmospheric chemists, unless they were interested meteorologically, would not have known particularly that there were polar stratospheric clouds."[25] And those who knew about polar stratospheric clouds, like Farman, would not have realized their significance for ozone depletion.

An important role of the assessment process is to bring specialists from diverse areas together, as the ozone assessments in the 1980s began to do. But even as they did, these specialists did not necessarily respect each other's knowledge and expertise. Farman explained:

> In those days, we all lived in our little cocoons of our professional work. I knew about clouds in the Antarctic, but I wasn't a chemist, and certainly the chemists had no idea what the real world was like. To them it was glass little tubes in laboratories, and that's got nothing to do with the free atmosphere either, in a sense. I mean the number of times in the early stages we had people come along and say, "Here are the new rates," and then later, well yes, their rates didn't have anything to do with what goes on. It's amazing how long that sort of thing takes to break down, all these built-in things you require in your operating.[26]

One group of chemists who knew about the Junge layer, Richard Cadle, Paul Crutzen, and Dieter Ehhalt, were an exception to this.[27] In a 1975 paper they presented the results of a computer model including a reaction of liquid water, carried by the sulfate particles, with nitrogen pentoxide (N_2O_5, a gas).[28] The model predicted that the heterogeneous reaction of liquid water and N_2O_5 could play a role only in the lower stratosphere, below approximately 25 km altitude. Since most ozone depletion was thought to occur at an altitude of about 40 km, where the sulfate particles would be very sparse, homogeneous reactions there would swamp the slower, less common heterogeneous reactions.

The results seemed to indicate that the heterogeneous component was unimportant. Crutzen explained: "Because the N_2O_5 plus water reaction on the surface of [sulfate] particles was supposed to be slow—which was wrong, but it was supposed to be slow—then the whole credibility that heterogeneous reactions could play a role in stratospheric chemistry was negated."[29] Moreover, sulfate particles would only be present in significant quantities after a major volcanic eruption. Therefore, Crutzen and his colleagues concluded that "heterogeneous ozone decomposition in the stratosphere is possible but not established as being important."[30] Others working on the topic agreed that "studies indicate a negligible contribution of the heterogeneous chlorine atom reactions compared to the Cl-CH_4 homogeneous reaction."[31]

In April 1982, when the Mexican volcano El Chichón erupted, sending large quantities of particles into the stratosphere, Sherwood Rowland reconsidered, thinking that chlorine nitrate ($ClNO_3$) or hydrochloric acid (HCl) from volcanoes might react with water or ozone on the surfaces of the volcanic particles. Working with Haruo Sato and others, he found in the laboratory that these reactions occurred almost instantaneously on any and every available surface.[32] Rowland explained:

> This was a small container and we repeated the experiment with the inert materials like paraffin—there are a set of inert materials that gas phase kinetics people have worked out over a period of time which they put . . . on the surface of the inside [to prevent heterogeneous reactions]—and we tried all of the known ones that made inert surfaces and got the same reaction rate. . . . It went as soon as you opened the stopcock. . . . Chlorine nitrate was a known compound, but one which is very hard to handle, because in a vacuum line if there were any trace of moisture, it immediately went to nitric acid and HCl.

Were there enough surfaces in the stratosphere to catalyze the reactions? Rowland still did not think so, believing at the time that "in the stratosphere, there are no surfaces."[33]

Meanwhile, Rowland's former postdoctoral fellow Mario Molina was also studying chlorine nitrate in the mid-1980s (having moved to the Jet

Propulsion Laboratory in Pasadena, California). Molina's work focused on reactions with HCl, not water. In a paper published one month after the ozone hole discovery, Molina and colleagues discussed experiments in which they remeasured the rate of the homogeneous reaction between HCl and $ClNO_3$, a reaction that could release Cl from its reservoirs. When some of their results showed a very rapid reaction, they explained these anomalous results as a laboratory artifact, writing, "The occurrence of this reaction has been noted by other investigators as well, with results that are also most likely explained by surface effects." Finding a likely culprit in Teflon surfaces in the laboratory apparatus, they concluded:

> Our experimental results do not rule out the occurrence of [this reaction] in the stratosphere as a heterogeneous process. However, given the relatively small collision rates with particulates . . . and considering that both reactants have to be on the surface, we feel that it is not likely for such a heterogeneous process to contribute significantly to the release of Cl from HCl.[34]

In hindsight, one might say that Molina and colleagues had actually demonstrated that chlorine reactions happen very quickly in the presence of surfaces to catalyze them, but that is not how they interpreted their results at the time. Rather, knowing of no such surfaces in the stratosphere—or at least knowing of none present in the quantities needed—they dismissed their own results. Rowland and colleagues had similarly recognized that chlorine nitrate hydrolysis would occur rapidly in the presence of surfaces, but again, knowing of no available surfaces, they dismissed the importance of the effect.

A few scientists did test for heterogeneous reactions on ice particles (rather than sulfates), but this work was fraught with difficulty too. In this case, the validity of the heterogeneous reaction of HCl and $ClNO_3$ would depend on the solubility of HCl in the ice. This was thought to be very low, based on laboratory measurements dating back to the late nineteenth century. After the Antarctic ozone hole discovery, chemists noted that the solubility measurements had been done on ice only just below freezing—nowhere near as cold as the ice particles in polar stratospheric clouds. When Molina and colleagues repeated the tests at temperatures approaching −78°C (the temperature of polar stratospheric clouds), the solubility was found to be significantly greater, which would facilitate the heterogeneous reaction.[35] But it did not occur to anyone to conduct experiments at such low temperatures—at least not until the discovery of the Antarctic ozone hole.

OZONE ASSESSMENTS IN THE UNITED STATES AND UNITED KINGDOM

By 1975 there was already enough concern about protecting the ozone layer for the first scientific assessment to be conducted. Over the next decade there were more than a dozen assessments, undertaken by both national governments and international organizations. In this section we consider several early ozone assessments sponsored by national governments: the US Department of Transportation's Climate Impacts Assessment Program (CIAP), 1971–1975; a two-part ozone assessment produced by the US NAS in 1976; and two reports by the UK Department of the Environment (see table 3.1).

TABLE 3.1. Some early assessments related to ozone depletion.

Ozone Report Title	*Year*	*Country and Institution*
Series of 5 monographs: *I: Panel on the Natural Stratosphere; II: Propulsion Effluents in the Stratosphere; III: The Stratosphere Perturbed by Propulsion Effluents; IV: The Natural and Radiatively Perturbed Troposphere; V: Impacts of Climatic Change on the Biosphere*	1975	US Department of Transportation, Climatic Impacts Assessment Program
Halocarbons: Effects on Stratospheric Ozone	1976	US National Academy of Sciences, Panel on Atmospheric Chemistry
Halocarbons: Environmental Effects of Chlorofluoromethane Release	1976	US National Academy of Sciences, Committee on Impacts of Stratospheric Change
[various reports]	1977–1985	United Nations Environment Programme Coordinating Committee on the Ozone Layer
Chlorofluorocarbons and Their Effect on Stratospheric Ozone, Pollution Paper no. 5	1976	UK Department of the Environment
Chlorofluorocarbons and Their Effect on Stratospheric Ozone (Second Report), Pollution Paper no. 15	1979	UK Department of the Environment: Part 1—Directorate of Air, Noise and Wastes; Part 2—Stratospheric Research Advisory Committee

The US Department of Transportation Climate Impacts Assessment Program (1975)

Sponsored by the US Department of Transportation, CIAP was formed in 1971 to evaluate the risks posed by the development of a large fleet of SST aircraft. CIAP involved work by over a thousand university, government, and industry scientists from 10 countries and contributed enormously to the state of knowledge of atmospheric science and modeling.[36] In 1975, CIAP scientists concluded that the exhaust produced by the proposed SST fleet would pose a serious threat to the ozone layer.

However, in January 1975, two months before the full report was made public, the credibility of CIAP was undermined by an executive summary (and accompanying press release) that seemed to suggest the opposite. The executive summary effectively changed the subject, dwelling mostly on the relatively negligible effects of a small and near-term projected SST fleet (30 or so aircraft), and downplaying the possible effects that scientists predicted of the long-term projected fleet of 200 or so aircraft. The effects of this large SST fleet were cast as preventable through future unspecified technology yet to be developed. The overall effect of the tone and wording of this executive summary, Parson noted, was to "suggest that CIAP's results had refuted, rather than largely confirmed, the initial ozone-depletion concerns" raised about SSTs—a suggestion that was entirely erroneous. Indeed, "A wire service report of the press conference made this misinterpretation explicit and was widely repeated, in some cases with scathing attacks on the scientists who had raised the alarm."[37]

Several scientists who had participated in CIAP came forward to object, noting that the executive summary had been rewritten by the CIAP manager, Alan J. Grobecker of the US Department of Transportation, and his staff, without the knowledge of the scientists and without distributing the draft for their consideration. The rewritten executive summary reordered, reworded, or omitted a number of the scientific conclusions and precise wordings that were used in the report that had already been written and approved by CIAP scientists. That report suggested that CFCs could damage the ozone layer.

US National Academy of Sciences 1976 Reports

In 1976, funded by NASA, the National Oceanic and Atmospheric Administration (NOAA), the Environmental Protection Agency (EPA), and the Federal Aviation Administration (FAA), the NAS produced a pair of reports. The first, by the academy's Panel on Atmospheric Chemistry, was called *Halocarbons: Effects on Stratospheric Ozone* and focused on science.[38] The second, produced by an ad hoc Committee on Impacts of Stratospheric Change (CISC), was called *Halocarbons: Environmental Effects of Chlorofluoromethane Release* and was charged with making policy recommendations based on the science presented in the first.[39]

Although the reports were presented separately, an executive summary included with the committee's report presented the conclusions of both the panel and the committee. Moreover, the chair of the panel also served on the committee, two other members served on both the panel and the committee, staff members in charge of both reports were the same, and the panel and the committee held some joint meetings. In the preface to the committee's report, its chair, the distinguished statistician John W. Tukey, explained the relationship between the charges of the committee and the panel:

> A panel . . . was established in April 1975 to study the question of whether the chlorofluoromethanes [the first class of CFCs examined as ozone-depleting substances] would destroy ozone, and if so, what the magnitude of the effect would be and what uncertainties were associated with such a prediction. The panel was also asked to identify critical research needs to reduce the uncertainties. The CISC meanwhile considered the question of biological and climatic effects of ozone reduction and the appropriate policy consequences of both our present knowledge and the knowledge we are likely to have in the future.[40]

The panel focused on Freon 11 and Freon 12 (the two CFCs first identified as ozone depleting) and concluded that if they continued to be released at then current rates, they would eventually cause 2%–20% depletion of the world's ozone layer, with the most likely steady-state value around 7%. However, they cautioned that a better understanding of the presence and action of chlorine nitrate ($ClNO_3$ or $ClONO_2$) in the stratosphere could change ozone depletion estimates significantly.[41]

This was not just a routine caveat; uncertainty over chlorine nitrate led to delaying the publication of the reports by several months. Chlorine nitrate, a reservoir compound that ties up both chlorine and nitrogen, was originally thought to dissociate quickly in sunlight; in the laboratory, chlorine nitrate broke up as quickly as it formed. If it behaved the same way in the stratosphere, then it would do little to prevent chlorine and nitrogen from depleting ozone. However, in early 1976 Rowland and colleagues presented work suggesting that chlorine nitrate might persist for some time under certain atmospheric conditions. The original ozone depletion calculations they reported in their 1974 paper were based on other scientists' published laboratory results regarding the behavior of chlorine nitrate. However, those fast laboratory reactions in which chlorine nitrate broke up as quickly as it formed were heterogeneous ones—occurring on the surfaces of laboratory apparatus—and it was not clear whether the same reactions, at the same speed, would happen in the stratosphere. Using an updated value for the expected lifetime of chlorine nitrate, Rowland and colleagues now found that it might persist long enough to significantly reduce the amount of ozone depletion caused by chlorine and nitrogen.

The academy delayed the report to work on this point. The best estimate the academy proffered—7% ozone depletion—took into account the most up-to-date information on chlorine nitrate, but the authors cautioned:

> If $ClONO_2$ proves to be less important [i.e., less active in tying up chlorine and nitrogen] than indicated by the present data, the ozone reduction could be larger than the stated values by a factor of up to 1.85.[42]

The basic science was still in flux, and scientists warned that ozone depletion might turn out to be twice as bad as their present best estimate.

In this unsettled state of scientific affairs, the committee seized on a statement by panel scientists that in two years' time the scientific community might be able to gain a better overall understanding of the key parameters controlling ozone in the stratosphere and "significantly reduce the possibility of subsequently finding an unidentified factor that has a major effect on predictions of ozone reduction by the CFMs." The panel also suggested that it would be able to improve the relevant models and reduce uncertainties overall. On that basis, the committee recommended delaying CFC regulation for two years.[43]

Presumably reflecting a difference of opinion and interpretation within the committee, the recommendation was presented in the executive summary in two similar but not identical ways. Recommendation 1 said:

> As soon as the inadequacies in the bases of present calculations are significantly reduced, for which no more than 2 years need be allowed, and provided that ultimate ozone reductions of more than a few percent then remain a major possibility, we recommend undertaking selective regulation of the uses and releases of CFMs on the basis of ozone reduction.[44]

Here it seems that the committee was recommending regulation within two years or perhaps sooner unless it was demonstrated that "ultimate ozone reductions of more than a few per cent" are not a "major possibility." Recommendation 6, however, seems to recommend against regulation, while suggesting that the question be revisited in "a year or two" when measurements would be better. It states:

> In view of the present inadequacies in the bases of our calculations, in view of the reduction in these inadequacies promised by ongoing measurement programs, and in view of the small changes in ozone reductions following from a year or two delay, we wish to recommend against [a] decision to regulate at this time.[45]

These recommendations are consistent in that both express the expectation that the "inadequacies in the bases of present calculations" would be reduced soon, neither suggests regulation now, and both suggest that a solid scientific basis for regulation will be available within two years. On the other hand, the tone of the recommendations is different. The first clearly recommends regulation after a short delay; the second is more ambiguous as to what the ultimate outcome will be.

These differences in tone and emphasis presumably reflect different levels of concern, different views of the uncertainties, and perhaps different attitudes toward regulation and precaution. At this point no one had actually observed ozone depletion; scientists were grappling with predictions, based on uncertain and incomplete information, about something that might prove to be very damaging but as yet had not actually done damage—at least so far as they knew.

In the context of what we know about the assessment process—wherein diverse individuals are trying to come to agreement on specific findings and recommendations—it is easy to conclude that the committee was grappling with how both they and policy makers should respond to the fact that substantive portions of the science were still in flux (but expected to be resolved soon). At the time, however, it was easy to see recommendation 1 as recommending regulation and recommendation 6 as declining to do so. Not surprisingly, advocates and opponents of regulation applied different portions of the text in support of their views. Supporters of regulation took the committee report to mean regulation would be needed in two years at most, while opponents (including CFC industry officials) took it to mean the science was so unsettled that we should wait at least two years before considering regulation. The media reflected this division. The day after the release of the reports the *New York Times* ran the headline "Scientists Back New Aerosol Curbs to Protect Ozone in Atmosphere," while the *Washington Times* headline claimed "Aerosol Ban Opposed by Science Unit."[46]

Science writer Sharon Roan has argued that the main reason for the "waffling" in the report was the panel's and the committee's discomfort with their explicit policy mandate.[47] Edward Parson concurs, noting that several national governments and "regulatory agencies resisted calls for immediate action by saying they were awaiting the Academy reports to act," putting considerable pressure on them to work wisely and well.[48] Panelist Harold Schiff later stated that the panel expected that regulatory decisions would be based on its findings, and this "made us a bit more cautious, perhaps, in phrasing and drawing conclusions because I think we were responsible enough to realize we were fooling with a fairly major industry."[49] Evidently, at least some of the scientists involved felt nervous, perhaps even intimidated, by the thought that their conclusions might on the one hand drive significant public policy and on the other invoke the wrath of a powerful industry.

Parson has judged the academy harshly for attempting to make policy recommendations, arguing that the reports "established a harmful model for scientific assessments and weakened the credibility of subsequent Academy reports on the issue."[50] But he ignores the fact that the committee was formed and asked to make policy recommendations by the White House Office of Science and Technology Policy. Given that the academy was created to respond to requests for advice from the federal government, it is hard to imagine a plausible scenario in which they would have declined this request or a plausible argument that they should have.

UK Department of the Environment: 1976 and 1979

In 1976 and 1979 the UK Department of the Environment published reports, both entitled *Chlorofluorocarbons and Their Effect on Stratospheric Ozone*. These are of particular interest because of the way they managed the relation between science and policy in contrast to the NAS reports.

These UK reports were fundamentally government policy documents based on assessments of the current science, and their structure reflected that. The lengthiest part of each report was an assessment of the science prepared by a committee of government, university, and industry scientists whose members were largely the same in both reports (in the 1976 report the scientific assessment was the "appendix," and in the 1979 report it had graduated in nomenclature to "Part 2"). Preceding the scientific assessment was a relatively short report written by government officials that summarized the scientific assessment and made policy recommendations. Preceding this in the report was a one-page foreword by the secretary of state for the environment declaring the position of the government of the day. The foreword to the 1976 report consisted of six paragraphs; the foreword to the 1979 report consisted of five numbered paragraphs.

The 1976 report was in part a response to a 1974 report by the independent Royal Commission on Environmental Pollution, which expressed strong concern about ozone depletion.[51] The appendix to the 1976 report drew on a British model that predicted 8% global ozone depletion after a hundred years. The main body of the report noted that if these models were correct, then action would "need to be taken long before the ozone reduction reaches any internationally agreed critical levels and indeed possibly before any measurable decrease can be detected at all."[52] It went on to emphasize uncertainties, to suggest that earlier concerns about building a fleet of SSTs had been "exaggerated," and to note the supposed benefits (as well as harms) to people of UV radiation. In the section titled "Conclusions and Recommendations" these government officials wrote:

> The consequence of the eventual 16% increase in UV radiation predicted assuming continued production of CFCs at 1973 rates is no worse than the increase currently experienced in moving from the north to the south of England. Such an increased level will not be reached for about 100 years and although regulatory action may be needed long before this, there appears to be no need for precipitate action.[53]

In the foreword to the report, Anthony Crosland, Labour Party politician and secretary of state for the environment, emphasized scientific uncertainty and the international dimensions of the problem. He wrote:

> The problem, if it exists, is not specific to the UK nor to Europe. It concerns all industrialized nations. If therefore there is a need for action, this can only be effective on an international basis.

He concluded that "in two or three years' time we shall be better informed to make decisions about the continued use of chlorofluorocarbons."[54] He added that in the interim, "prudence demands that attempts should be made to seek alternatives which meet the needs of the public and of industry but which do not pose the same potential threat to the stratosphere."[55]

By 1979, the UK secretary of state for the environment was Michael Heseltine, a prominent figure in Margaret Thatcher's new Conservative government, whose main interest in this role was in promoting a policy that allowed residents of public housing to buy their apartments ("right to buy"). His foreword to the 1979 report repeated much of the substance and even language of Crossland's 1976 foreword but in a more skeptical tone:

> This theory [that CFCs cause ozone depletion with "possibly adverse consequences for human health"] has been the subject of controversy since it was first propounded in 1974. . . . The validity of the hypothesis is still in doubt but if it is right and if emissions were to continue at the earlier levels the threat would increase slightly with each passing year. . . . The problem, if it exists, is not specific to the UK or to Europe—it is a global issue.[56]

While the ministers' forewords had tread water or even retreated from 1976 to 1979, the science had continued to develop, and this was reflected in the scientific assessment that constituted part 2 of the report. Indeed, chapter 1 of part 2 is called "The Chlorofluorocarbon Problem." While allowing that more work was needed, particularly more observations of the stratosphere, it emphasized the general agreement in the scientific community that should CFC emissions continue, ozone depletion would occur. There was also general agreement as to the range of expected depletion—5%–20%—although the report did point out that consensus did not equate to truth, and "the fact that a number of independent workers who used these same assumptions calculate eventual ozone depletion in the same range . . . does not prove that the true figure would lie in this range."[57] Still, the overall conclusion was clear enough: ozone depletion was a real concern, and there were "no marked differences between results derived by modelers in the UK group and those from overseas research centres."[58]

The much shorter policy portion of the report (only 26 pages, compared to over 200 pages for the science) written by UK Department of the Environment officials, concluded that the uncertainties in the scientific findings, particularly the "discrepancies between model calculations and measurements[,] brings into question the validity of the models presently used to predict ozone perturbations," suggesting that these discrepancies indicated a "lack of understanding of stratospheric processes."[59] The authors also suggested that the problem might correct itself without government intervention; given how cautious people were starting to feel about CFCs, especially in nonessential aerosol propellant uses, "The amount of CFCs used in aerosols in the UK will decline to about 70% of the 1976 volume over the next three years," a reduction that "appears to be adequate pending further research." They concluded that "strict regulation is not warranted at present."[60]

In these early reports, American and British scientists were in broad agreement about both the risks of ozone depletion and the scientific uncertainties, although their tone was sometimes different. Even the positions of their governments were not as far apart as sometimes supposed. The British government was committed to supersonic transport, having launched regularly scheduled flights of the British/French Concorde in 1976. However, elements of the US government also supported SSTs during this period. In 1978 an American carrier began flying the Concorde from Dallas to Washington, and NASA continued to fund SST research throughout the 1970s.[61]

What was profoundly different was the background understandings of the relationships between the science and policy in these reports. The British reports were fundamentally policy documents issuing from the government of the day, and officials made statements in these reports that went beyond and in some cases seemed antithetical to what the scientists were saying in the reports. The American reports were seen as fundamentally scientific assessments or an explicit request for scientists to make policy recommendations. When US government officials were viewed as compromising or distorting what the scientists were saying, this provoked controversy and even outrage.

INTERNATIONAL OZONE ASSESSMENTS

In 1977 the United Nations Environment Programme (UNEP) formed the first international committee to address ozone depletion. The Coordinating Committee on the Ozone Layer (CCOL, which remained in existence until 1986) was charged with a range of tasks, including the coordination of and reporting on global scientific research. After its initial meeting, members consisted of all interested countries and organizations involved in ozone-related research. Nongovernmental organizations not conducting their own research could not participate, but the CFC industry–supported Chemical Manufacturers Association's Fluorocarbon Panel did.[62] The process consisted of weeklong annual meetings, during which assessment reports would synthesize scientific research from around the world.

There was no agreement within CCOL as to whether it should examine the socioeconomic aspects of ozone depletion, such as projected CFC emission rates, replacement chemicals and technologies, and policy options. In the first report the committee addressed only the physical science aspects. The next two reports made limited policy-relevant statements, but CCOL was soon overshadowed by other institutions and assessments.

Robert Watson, who would emerge as a major figure in international ozone assessment and chair of the Intergovernmental Panel on Climate Change, has said that the time allowed was too short to permit a comprehensive review, the budgets too small, and the people involved chosen too bureaucratically without sufficient regard to scientific expertise:

> Each country that had research in ozone depletion was allowed to send one or two people to a meeting in Nairobi once every couple of years, roughly. . . . At the same time we would write a very, very brief assessment, all in a week. So totally it'd have to be superficial, basically, compared to what was needed. And not every country sent the right person. I think because EPA was the normal link to UNEP for the US, the delegation was actually headed by a guy called something like Wisener or Whizner [Herb Wiser]. . . . He headed the delegation, and the US EPA budget was peanuts compared to the NASA/NOAA budget. After a couple of these, I got put on; I said, "This is bloody ridiculous. You've almost got no budget, and you don't know what the hell you're talking about." (I said it a bit more kindly, but not much more.) . . . He didn't have the budget to do it properly. . . . These little reports would be written in a one-week period, which is not the way to do them. Most of them were government bureaucrats going [to the meetings], of course (which I was by then, you could argue), far from getting the best scientists in the world.[63]

Watson received a PhD in chemistry at the University of London in 1973, worked as a scientist at the Jet Propulsion Laboratory in Pasadena, California, from 1976 to 1987, and from 1980 to 1987 also served as acting program scientist at NASA. In the beginning, Watson's research, measuring the rates at which chlorine, bromine, and fluorine atoms react with ozone in the laboratory, did not seem to have immediate application to the real world. However, he soon realized that it had huge relevance in a world facing the threat of ozone depletion.[64] As the specter of ozone depletion came to scientific attention, Watson and others realized that more work on chemical reaction rates was badly needed.

Watson believed that modelers needed a unified standard from which to choose the best reaction rates to put into their models. This would help to ensure that the most accurate and up-to-date scientific information was used in the computer models and permit model comparison. Otherwise, if different models produced different forecasts of ozone depletion, it would be hard to discern the sources of disagreement.

In the late 1970s Watson and others formed the Committee on Data for Science and Technology Task Group on Chemical Kinetics (CODATA), initially part of the US National Bureau of Standards.[65] Through laboratory measurements they would establish and publish the best and most current rates for various chemical reactions relevant to ozone depletion, along with uncertainty estimates for each reaction.

Watson was also involved with the international standardization of balloon-borne sounding systems, which measured ozone as well as aircraft-based measurements. It was often difficult to discern whether variation in results indicated actual fluctuations; were the result of having been measured at different times, places, and altitudes; or just reflected instrumental variability. According to Watson,

> [I] pulled a whole bunch of these balloon and aircraft intercomparison campaigns together where you put all the instruments measuring the same bloody thing at the same time. So you could see instrument-to-instrument comparison, so when they measured it then in different parts of the world, you could tell whether it was the instrument variability or was there really a change. . . . [These] were the things that sort of led to us having some confidence in—about what we were talking about.[66]

Watson believed that large-scale international assessments could perform a similar role. Rather than policy makers from each country organizing their own assessments and having to decide which of several available assessments provided the best summary of the current science and most reliable predictions of the future, they could all refer to a single assessment that had been produced with broad participation. In a world of multiple assessments from different countries and institutions, it was easy for those who opposed CFC regulation to seize on the differences among them as proof that the science was unsettled and regulatory decisions should therefore wait. A single international assessment would make this harder to do.

Meanwhile, Watson was working on US ozone assessments. The US Clean Air Act (1977) required NASA, along with FAA and NOAA, to perform biannual assessments of ozone depletion. For this purpose, NASA formed the Upper Atmosphere Research Program; Watson became its head in 1980, having organized the first two (1977 and 1979) assessments. While planning for the third NASA assessment, Watson approached fellow CODATA Task Group member Adrian Tuck. Then at the UK Meteorological Office, Tuck was a panelist on the 1979 British ozone assessment. Tuck declined to participate in Watson's assessment. According to Watson,

> He [Tuck] said, "Oh, I'd love to, but it's a US government assessment. I clearly can't be part of it because if indeed it comes out with some conclusions that are inconsistent with the UK position it would be very awkward saying, 'Look, this

> British scientist working for the British government . . . is part of an American assessment.'" I said, "This is just silly, Adrian. We've got to get past these national assessments."[67]

Tuck tells the story in a similar fashion: "I said [to Watson], 'We're going to have this international head-butting until we have international assessments.' That's how the WMO assessment started."[68]

Watson didn't just want the particular expertise of Adrian Tuck in the upcoming ozone assessment; he wanted to increase the authority of the report and the credibility of the process by involving scientists from around the globe:

> The whole reason for doing the first [international, i.e., 1981] assessment was to get one single assessment done. That was the whole rationale. . . . That effectively you need one assessment, but it must be open and transparent and involve experts from all countries. Otherwise it isn't open and transparent and you can't expect other people to use it. The whole philosophy of going to an international assessment was bringing everyone together so there weren't five reports or six reports in two years. That was the whole philosophy.[69]

The World Meteorological Organization (1981)

Soon after his conversation with Tuck, Watson approached Narasimhan Sundararaman, a colleague who had just been placed in charge of the ozone program at the World Meteorological Organization (WMO), to enlist his support.[70] Sundararaman agreed that the WMO would co-sponsor the assessment if NASA provided the funding. The result—*The Stratosphere 1981: Theory and Measurements*—was sponsored by NASA, NOAA, FAA, and WMO. Seventy-five percent of its authors came from outside the United States, including Australia, Belgium, Canada, China, France, Germany, Italy, Japan, Norway, Switzerland, and the United Kingdom (see table 3.2).

The Stratosphere 1981 was a single-volume report with a preface and an introduction but no executive summary. The authors divided themselves into working groups corresponding to the various chapters and appendixes of the assessment. The working groups developed "position papers" and then held five "pre-meetings" to begin to work these position papers into a coherent whole before convening for a final five-day workshop to produce a finished product. In the report preface, they explained their process:

> Prominent scientists active and expert in stratospheric and mesospheric studies were asked to prepare position papers to cover selected topics of research. In addition to the prepared position papers, pre-meetings . . . were held to

TABLE 3.2. International ozone assessments sponsored by the World Meteorological Organization (among other bodies).

Publication Date	*Title of Assessment*	*Chair or Cochairs*	*Sponsoring Organizations*
1982	*The Stratosphere 1981: Theory and Measurements*	Cochairmen Robert D. Hudson, Rumen D. Bojkov; executive council: Marvin Geller, Edith I. Reed, Richard S. Stolarski, Narasimhan Sundararman, Robert T. Watson	WMO, UNEP, NASA, FAA, NOAA
1986	*Atmospheric Ozone 1985: Assessment of Our Understanding of the Processes Controlling Its Present Distribution and Change* (3 vols.) (unofficially known as the Blue Books)	Overall chairman Robert T. Watson	NASA, FAA, NOAA, UNEP, WMO, EC, BMFT
1990	*Report of the International Ozone Trends Panel 1988* (2 vols. and separate executive summary)	Chair Robert T. Watson; vice chair Richard Stolarski	NASA, NOAA, FAA, WMO, UNEP
1990	*Scientific Assessment of Stratospheric Ozone: 1989* (2 vols.)	Assessment cochairs Daniel L. Albritton and Robert T. Watson	NASA, DOE, NOAA, UNEP, WMO[1]
1992	*Scientific Assessment of Ozone Depletion: 1991*	Assessment cochairs Daniel L. Albritton and Robert T. Watson	NASA, NOAA, DOE, UNEP, WMO[1]
February 1995	*Scientific Assessment of Ozone Depletion: 1994*	Assessment cochairs Daniel L. Albritton, Robert T. Watson, and Piet Aucamp	NOAA, NASA, UNEP, WMO[1]
February 1999	*Scientific Assessment of Ozone Depletion: 1998*	Assessment cochairs Daniel L. Albritton, Pieter Aucamp, Gérard Mégie, and Robert T. Watson	NOAA, NASA, UNEP, WMO, EC[1]

(*continued*)

TABLE 3.2. *(continued)*

Publication Date	*Title of Assessment*	*Chair or Cochairs*	*Sponsoring Organizations*
March 2003	*Scientific Assessment of Ozone Depletion: 2002*	Assessment cochairs Ayité-Lô Nohende Ajavon, Daniel L. Albritton, Gérard Mégie, and Robert T. Watson	NOAA, NASA, UNEP, WMO, EC[I]
February 2007	*Scientific Assessment of Ozone Depletion: 2006*	Assessment cochairs Ayité-Lô Nohende Ajavon, Daniel L. Albritton, and Robert T. Watson	NOAA, NASA, UNEP, WMO, EC[I]
March 2011	*Scientific Assessment of Ozone Depletion: 2010*	Assessment cochairs Ayité-Lô Nohende Ajavon, Paul Newman, John Pyle, and A. R. Ravishankara	NOAA, NASA, UNEP, WMO, EC[I]

Note: Publication dates, where available, are from the sources themselves and/or from WorldCat. Organizations are listed in the order in which they are presented in the assessments themselves. Abbreviations are as follows:
BMFT: Bundesministerium für Forschung und Technologie (Germany)
DOE: Department of the Environment (UK)
EC: European Commission (or Commission of the European Communities)
FAA: Federal Aviation Administration (US)
NASA: National Aeronautics and Space Administration (US)
NOAA: National Oceanic and Atmospheric Administration (US)
UNEP: United Nations Environment Programme
WMO: World Meteorological Organization
[I]Assessment produced under the official mandate of the Montreal Protocol.

assemble some of this material into a cohesive report. The position papers and these prepared reports were furnished to the participants before they came to the workshop. An important aspect of this work was that the modeling groups agreed on a common set of scenarios to study so that realistic model comparisons could be made at the workshop.[71]

The CCOL reports were also international, but there were several significant differences between them and this new international assessment. First, all the participants in producing *The Stratosphere 1981* were scientists active and expert in disciplines relevant to stratospheric science and ozone depletion—no bureaucrats or policy makers were involved. Second, participants met much more often; CCOL members had to produce their entire report during a single weeklong meeting. Third, a significantly greater amount of preparatory work went into the 1981 assessment, in the form of the "position papers" prepared by some but read by all, before any of these meetings took place. Like the CCOL reports, *The Stratosphere 1981* made no policy recommendations. The authors also stressed in the preface that they had not attempted to produce a "consensus document." Wherever "more than one conclusion could be maintained by the scientific data, then both of these conclusions have been quoted."[72]

The "Blue Books" (1986)

The 1981 effort set a template for future international ozone assessments, the first of which, chaired by Bob Watson and sponsored by NASA, WMO, UNEP, and various other national and international organizations, was published in 1986. The authors included over 150 scientists from 11 countries. The resulting three-volume report, *Atmospheric Ozone 1985: Assessment of Our Understanding of the Processes Controlling Its Present Distribution and Change* (known informally as the Blue Books for the color of the covers) is generally regarded both as the first truly international ozone assessment and as an unparalleled success in terms of its influence on global policy.[73]

The process followed was similar to that used to produce the 1981 report but expanded. It began with a weeklong meeting (in Germany, in June 1984) of 80 scientists, who reviewed the current state of scientific knowledge and chose working group chairs, and thereby chapter lead authors. At the next meeting, additional authors for the various groups were chosen. The working groups met periodically over the next year. More than 30 chapter meetings took place in all. Draft chapters were presented for review at a weeklong workshop in July 1985 in Les Diablerets, Switzerland. The drafts were reviewed not only by other attending chapter chairs but also by a panel of outside scientific experts. Final versions were submitted in November and December of 1985, and the report was published in 1986.[74]

Like *The Stratosphere 1981*, the Blue Books did not have an executive summary. There was, however, a 25-page "Introduction and Science Summary" that presented (in a series of numbered lists ranging from half a page to two pages) the "Major Conclusions and Recommendations" from each chapter of the report. Its format consisted of introducing the topics each chapter was going to cover, rather than presenting the major conclusions from the assessment in a concise and easy-to-read fashion.

In the Blue Books, the scientists had managed to produce something like a textbook on a subject for which there was no actual textbook and was not yet taught in most schools and universities, but about which people outside of the scientific community needed to know. As Bulgarian physicist (and author) Rumen Bojkov explained, "The blue books really read like a textbook, which you can use to get a comprehensive introduction to the field." In addition to summarizing the prevailing scientific view of the time, it also "provid[ed] the detailed background necessary to support the view."[75]

While steering clear of explicit policy questions, the Blue Books did present some sobering conclusions. Among these were the results from one- and two-dimensional models that were programmed with various values for CFC emissions and ozone-depleting reactions, results that appeared far more worrisome than the 7% column depletion suggested by the 1976 NAS panel report. Consider this statement highlighting local depletion at one altitude rather than the column depletion that averages over all altitudes: "All models with all scenarios predict that continued release of CFCs 11 and 12 at the 1980 rate will reduce local ozone at 40 km by ~40% or more."[76] The implications of this conclusion were easily grasped. Parson writes that "the authority of the [Blue Books] assessment was so strong, its involvement of the top researchers in the field so nearly total" that it was able to avoid

> being attacked and . . . diminishing its authoritative standing. The most important such statement—that significant ozone loss was likely under CFC growth . . .—became the principal means by which the 1985 assessment subsequently exercised decisive policy influence.[77]

With the Blue Books, scientists may have achieved political credibility and scientific authority, but something was still missing. They had discussed serious ozone depletion as something that would occur in the future, but they soon discovered that it was already under way. Just as the Blue Books were nearing completion, British scientists announced their discovery of the ozone hole. Scientific expeditions were soon mounted to study this phenomenon and determine if it was indeed caused by chlorine from anthropogenic CFCs. One 1986 expedition suggested that this was the case. A second expedition the following year confirmed it. The question that then emerged was whether this was a global phenomenon or one restricted to the unique atmospheric conditions of the Antarctic. This became a matter of intense scientific research.

The Report of the International Ozone Trends Panel *(1988–1990)*

The International Ozone Trends Panel (OTP) was formed in 1986, under Bob Watson's leadership and sponsorship from NASA, NOAA, FAA, WMO, and UNEP. The report's executive summary was released in 1988, although the full two-volume report was not released until 1990.[78] The report might have simply been an expansion and extension of what had been done before, except for the urgency introduced by the discovery of the ozone hole.

The 1985 discovery, with its confirmation in 1986 and 1987, was highly publicized, garnering huge media attention and generating great public concern. But evidence of serious ozone depletion was not entirely new. In 1981, and again in 1985, NASA scientist Don Heath had announced the detection of significant global ozone depletion by the Nimbus-7 NASA satellite via solar backscatter ultraviolet measurements.[79] However, various scientists dismissed these results as spurious. It was well known that the diffuser plate on the ozone measuring apparatus was slowly deteriorating over time.[80] This deterioration introduced an artificial reduction in the amount of ozone detected by the satellite instruments—a reduction that had to be compensated for in any calculations made with those measurements. If this degradation were not properly taken into account, then ozone levels would seem to be declining even when they were not.

Meanwhile, Rowland had been reexamining the history of ozone measurements from a station in the Dobson network in Arosa, Switzerland.[81] He had noticed a possible decline in ozone levels at that station, as well as at two other stations located at high latitudes in the northern hemisphere. Wanting to find out if this was a statistical anomaly or a genuine signal, he asked Neil Harris, at that time a graduate student working on his PhD under Rowland's supervision, to analyze the data. Rowland recounted instructing Harris:

> When I look at the ozone results from Arosa, it looks to me as though it's going down in the last few years. But everybody who is looking at it says no, the statistics shows there is not [an] effect, or that it is actually increasing. What I want you to do is to figure out where my eyes disagree with their statistics.[82]

Harris began this work around the same time the OTP was forming, and Watson decided to include Rowland and Harris's work on the ground-based ozone data as part of the analysis. Rowland recalled a discussion at a meeting of the American Geophysical Union in San Francisco in 1985: "[Watson] put together the whole business [i.e., Heath's satellite data plus their reevaluation of the ground-based data] and turned to me at one point and said, 'Since you've been evaluating the ground-based occurrence, why don't you take that subcommittee?' "[83]

As its name suggests, the Ozone Trends Panel was charged with analyzing recent satellite- and ground-based reports of downward trends in global ozone levels (the panel eventually also took on the task of analyzing the reports of severe Antarctic ozone loss). The panel set out on a massive campaign to reanalyze all of the relevant data.

The panel's report concluded that while Heath's results overestimated observed ozone depletion, other studies had underestimated it. Taken together, the studies suggested a small but observable downward trend in global ozone levels. One reason this trend had been difficult to detect, as Harris and Rowland now realized, was that scientists working with the ground-based measurements had been averaging their data, annually and globally, and giving more weight to the less variable (and therefore presumably more reliable) data from the summer months. However, when the data were analyzed on a month-by-month and station-by-station basis, significant nonpolar ozone depletion could be detected, particularly in the austral winter. Watson explained:

> What people had actually done [prior to the Ozone Trends Panel] was they took all the ozone data from all the stations across the world and they did an annual average of it, and said, "Is ozone decreasing globally?" No real signal. Why would you bring all the stations together at one time? Well, you've got more data, and if there's a small trend, the more chance of seeing it, the more data you got. But what it masked completely . . . was that the ozone depletion varies with latitude. . . . So what Sherry Rowland had noticed, and some others, was if you looked to some of the individual stations, they seemed to be showing ozone depletion, but when you put all of it together, they didn't seem to be showing depletion. But . . . you know, you can't rely on a single data [*sic*] and a single station; it's noisy, etc.[84]

Because the effect varied with latitude and season, when the data were grouped, the effect was masked.

The OTP undertook an original in-depth analysis and comparison of both ground-based and satellite ozone readings, applying new statistical methods to raw data in a way that hadn't been done before. It did so in response to scientists' conviction of need, a concern that something important had perhaps been missed in previous work. The executive summary noted:

> This report is different from most previous national and international scientific reviews in that we did not simply review the published literature, but performed

> a critical reanalysis and interpretation of nearly all ground-based and satellite data for total column and vertical profiles of ozone. To aid in the interpretation of the results of this re-analysis we also performed a series of theoretical calculations for comparison with the reanalyzed ozone data. In addition, a uniform error analysis was applied to all the data sets reviewed that contained information on the vertical ozone distribution.[85]

The result of this research was the finding that ozone depletion had occurred not just over the South Pole but globally. Though the global levels of depletion were not as dramatic as that over Antarctica, in some ways the results were more alarming, since few people lived in Antarctica. For the first time, significant ozone depletion had been detected over the highly populated Northern Hemisphere. Furthermore, the OTP directly attributed the depletion to atmospheric concentrations of ozone-depleting substances that "continue to increase on a global scale because of human activities."[86]

The reanalysis of the data was crucial for detecting the depletion, and crucial to the reanalysis was the inclusion of Sherwood Rowland, who had looked closely at the Arosa data and decided to follow up on it. Although Rowland was a key figure in ozone research—he would later share the Nobel Prize for this work—he was not often asked to participate in ozone assessments. Rowland attributed this to his early and outspoken support for CFC regulation: "I was always eliminated from being a panelist because I had expressed my opinion," he concluded.[87] Watson agreed, at least as far as the prospect of Rowland taking a leadership role:

> In my opinion, and it is a personal opinion, you couldn't have Sherry [Rowland] ever chair one of these international assessments. He was already on record on what he would do with the CFCs [from repeated congressional testimonies, interviews, and guest lectures], so therefore those that actually had a counterview to Sherry, if you'd have allowed Sherry to chair an international assessment, people would have said he was just talking [*sic*] the results to fit his preconceived ideas.[88]

Why, then, did Watson ask Rowland to be a lead author on the Ozone Trends Panel report? Despite his reservations, Watson's stated reason for involving Rowland is straightforward: "I did get him involved in a very visible way in the Ozone Trends Panel. He is a world-class scientist."[89] Watson evidently judged the value of Rowland's contribution, as a lead author if not a chair, to outweigh the "taint" of his known policy leanings.

Whatever some people may have thought about Rowland's public statements on ozone regulation, the OTP was broadly viewed as extremely important, credible, and effective, and its conclusions had a major impact on policy. It rode a wave of success created by the Blue Books as well as a wave of urgency and public concern. As a result, industry and policy makers took its conclusions very seriously.[90]

On March 18, 1988, three days after the Ozone Trends Panel released its executive summary, DuPont (the leading US manufacturer of CFCs) announced that it would stop manufacturing the chemicals within 10 years—an announcement that stood in stark contrast to DuPont CEO Richard Heckert's statement only two weeks before (on March 4, 1988) that there was as yet no evidence that CFCs posed a significant threat.[91] Although one DuPont scientist, Mack McFarland, had served on the OTP, panel members were sworn to secrecy, and there is no specific evidence that McFarland revealed the findings of the OTP to DuPont executives before the March 15 press release. (If he had, Heckert might have rethought his March 4 comments.) On the other hand, the OTP announcement did not come as a complete surprise. DuPont had scientists of its own working on the issue, who were aware of the progress in this area. Moreover, although OTP members were not free to discuss the panel's findings in public before the March release of the executive summary, their results were available months earlier in various reports by NASA. The results were also presented at the Dahlem Workshop on Antarctic ozone, held the previous November in Berlin. Still, the release of the OTP executive summary was public in a way that this other work was not. That its release was followed so closely by DuPont's decision confirmed scientists' views that the assessment was a success. This helped to make the structure they had used in the OTP a model for future assessments.

Ongoing Assessments under the Montreal Protocol

The Montreal Protocol on Substances That Deplete the Ozone Layer (a protocol promulgated by the Vienna Convention for the Protection of the Ozone Layer) was agreed to on September 16, 1987, and entered into force on January 1, 1989. It is an international treaty designed to protect the ozone layer by phasing out the production of numerous substances that are responsible for ozone depletion. It has been revised by amendment five times and by adjustment thirteen times. Former United Nations secretary general Kofi Annan called it "perhaps the single most successful international agreement to date."[92]

The Montreal Protocol requires periodic assessments of ozone science as a basis for future adjustments and amendments to the protocol. Annex VI of the protocol, "Assessment and Review of Control Measures," holds that

> beginning in 1990, and at least every four years thereafter, the Parties shall assess the control measures provided for in Article 2 on the basis of available scientific, environmental, technical and economic information. At least one year before each assessment, the Parties shall convene appropriate panels of experts qualified in the fields mentioned and determine the composition and terms of reference of any such panels. Within one year of being convened, the panels will report their conclusions, through the secretariat, to the Parties.[93]

Following the exact wording of annex VI, four panels were created: scientific, environmental, technical, and economic; later the third and fourth were merged to form the Technical and Economic Assessment Panel. The official duties of the Panel for Scientific Assessment included selecting internationally recognized experts chosen both for expertise and geographical balance; preparing a report reviewing new scientific information; and preparing an executive summary of the scientific report "written in a style understandable and useful to policy makers."[94]

Two years earlier, UNEP had established four panels to "review the current scientific, environmental, technical, and economic information relative to possible amendments of the Montreal Protocol"[95] as part of the pending US ozone assessments sponsored by NASA and spearheaded by Watson, then director and chief scientist of the Science Division of NASA's Mission to Planet Earth. Watson recalled a conversation with Mostafa Tolba, then the executive director of UNEP and heavily involved in the Montreal Protocol negotiations. Watson argued that the process he and others had begun on a national scale, with the NASA panels and assessments, should be broadened to an international scale:

> "Look . . . we've [i.e., NASA and a group of lead authors selected by Watson] already started . . . I've already got the authors lined up; I've already agreed the script." "Not a problem, Bob. . . . I'll say this looks like exactly what we need and we'll get it embedded with no changes." So that actually was really the way the thing got started.[96]

In this way, the panels that had been formed, the group that Watson had assembled under NASA auspices, and the chapter outlines they had already agreed upon became the framework not only for the next NASA assessments but also for the first international assessments under the Montreal Protocol. The duties of the four assessment panels were officially defined in annex VI to the Protocol, and Watson was asked to chair the assessment—in essence to continue officially and internationally in the role he had already unofficially taken on.

Scientific Assessment of Stratospheric Ozone: 1989, the report of a panel cochaired by Watson and Dan Albritton, a NOAA scientist, contained four chapters: "Polar Ozone," "Global Trends," "Theoretical Predictions," and "Halocarbon Ozone Depletion and Global Warming Potentials."[97] It had an executive summary, released separately and also published as part of the assessment. Following the mandates of the protocol, international scientific assessments have followed every four years, now under a standardized format and title, *Scientific Assessment of Ozone Depletion*.[98]

Beginning in 1994, it had a new section: "Common Questions about Ozone," and in 1998 "Frequently Asked Questions about Ozone" was added and then renamed and expanded in 2002 into "Twenty Questions and Answers about the Ozone Layer." Since 2002, the lead authors on these "Twenty Questions" have been US physicist David Fahey of NOAA and Swiss physicist Michaela Hegglin. In a 2009 interview, Fahey said that he was asked to do the Twenty Questions section by Dan Albritton: the official purpose was to answer the questions an educated person might ask if he or she knew nothing about ozone. Fahey agreed that this was needed: most available ozone information was either high science or at a grade-school level. However, another implicit but understood function of this section was to address "stupid" questions, erroneous views, and deliberate disinformation: Weren't CFCs too heavy to rise up to the stratosphere? Wasn't stratospheric ozone actually being damaged by chemicals from volcanoes?[99] Watson agreed, joking:

> What we wanted to call it was "Twenty Dumb Questions" . . . Aren't . . . volcanoes doing it? Aren't the fluorocarbons too heavy so they can't get into the atmosphere? They're really dumb questions. . . . But it turned out to be very policy-relevant and very useful.[100]

These were not just "dumb" questions. After all, why would an ordinary person even think that volcanoes caused ozone depletion or have any idea how heavy (or light) CFCs were? In fact, industry groups and think tanks had promoted these erroneous views, challenging the scientific evidence for political, economic, and ideological reasons.[101] By including a corrective section, scientists were (albeit implicitly) addressing these disinformation campaigns.

THE EVOLUTION OF OZONE ASSESSMENTS

The treatment of some important themes has changed as ozone assessments have evolved.

Participation

From the beginning, extrascientific considerations structured the choice of participants in ozone assessments. Communication skills, the ability to work in a team, and geographic balance all came to the fore. In time, it became generally accepted that lead authors not only had to be recognized experts in their respective scientific fields but must also possess strong communication and team-building skills.[102] It also became conventional wisdom that chapter teams needed balanced representation of men and women and of scientists from developed and developing countries.[103]

The issue of real or perceived bias also influenced who was invited to participate, but in interesting and not necessarily obvious or consistent ways. Sherwood Rowland was considered problematic because of his public stance on the need for stringent CFC regulation. It was not that his colleagues feared that he would bias the assessment. On the contrary, Rowland was widely admired by his colleagues for his judicious manner and even temperament. Rather, the worry was that anyone who wanted to dismiss the conclusions of a particular assessment would use Rowland's participation as an excuse to do so. Watson explained:

> You've got to get these assessments owned by everybody. Those on Sherry's [i.e., Rowland's] side would say, "Of course, this is great. Sherry chaired it—of course I believe it." Those who opposed it would say, "Damn it, we already know where he's coming from." So one has to be very careful.[104]

Yet sometimes participants—those from industry in particular—were involved precisely because they were seen as representing interest groups that had a stake in the outcome of the assessment. In some cases they were included to counter potential claims of anti-industry bias and to provide an information conduit to industry. Mack McFarland, for example, is a chemical physicist and was a contributing author to all the international ozone assessments beginning with the Blue Books. He was also a chief scientist for DuPont—one of the world's leading manufacturers of CFCs. McFarland reported that several people had told him that "[I am] valuable on these things because I taught the decision makers in industry."[105]

Watson's own role as author has also been affected by such considerations: he did not participate in the 2006 assessment but was asked by the other cochairs to let them keep his name on it in order to stave off potential political interference, lest it appear that the process was in flux and vulnerable to manipulation:

> My name is at the front of this one [2006], and I'm embarrassed because I actually didn't do much on this one. . . . I did almost nothing and I actually told them they should take my name off because I was embarrassed by that. But there was a political reason not to change while the [George W.] Bush administration was there. If it seemed to be redesigned, the Bush administration might try to put someone else on there that may not have been the right person. So they said, "Bob, even if you don't do anything, just allow us to use your name."[106]

John Pyle, a member of the steering committee on the 2006 assessment, confirmed that cochair Dan Albritton asked Watson to "stick around . . . because we don't know who might replace you."[107]

Finally, the issue of real or perceived bias has influenced lead authors' views on what type of person is the best person to participate in an assessment: an expert, yes, but one not too vested in promoting his own point of view. John Pyle, a modeler and chemist with the University of Cambridge, who has been a lead author on both ozone and Intergovernmental Panel on Climate Change (IPCC) assessments and was a cochair on the most recent ozone assessment (*Scientific Assessment of Ozone Depletion: 2010*), suggested that some scientists are too focused on promoting their own research and interpretations: "They want to push their particular agenda, push their research agenda or whatever: 'Here's my bit of work. I want to get it in the assessments.' "[108]

Selection criteria are weighted heavily in favor of scientists who are viewed as not too invested in a particular position—and who therefore will be open-minded, flexible, and willing to learn—but this can conflict with the very notion of expertise, since an expert often has a good deal invested in his or her expertise. As Bob Watson explained,

> You want the people to be highly knowledgeable, but you often did not want the person that may have published the most results in that area because they might conceivably only promote their own work. So the choice of lead authors is rather crucial: highly knowledgeable, but not necessarily the authors that have done the most work in that field because otherwise they may be slanted or biased to their own work. And you actually want people to defend their own work but be open to [the idea that] it may not be right.[109]

Similarly, Guy Brasseur, the director of the Climate Service Center in Germany, noted:

> Some of the authors are very fair, while others [might act like] "Oh, I'm going to report on my stuff and show what I did since I'm the best." OK, sometimes there

> might be disagreement over a very, very delicate issue. Let's say some people were talking about trends. One person measured a trend going up; another person measured no trend. Then the report needs to say something about that. So the report would say, "We think there's a trend. We think the other observations are wrong for that and that reason." The guy would come to the meeting and say, "Wait a minute. You're wrong! I can show you." So sometimes there's a fight.[110]

Generally speaking, scientists are not paid to participate in assessments, yet many willingly agree to do so. Most proffer that they benefit professionally, qua scientists, because participating in an assessment is an efficient means to get a view of a scientific domain that is both broad and deep, which can be helpful to both one's research and one's position in the research community. Several ozone scientists describe assessment participation as the opportunity to get "up to date over a wide range of science," particularly on "prepublication" research.[111] Pyle said, "It's a good way of keeping in touch with everything that's going on. Things are synthesized for you."[112]

Tony Cox, a member of the CODATA panels with Watson and a participant in all the international ozone assessments except 2010, suggested that the assessment process is important in stimulating new knowledge: "Whenever you get experts together, grappling with a problem [in] which everybody recognizes we don't know what's going on here . . . people stimulate each other and learn about the thoughts of others." The ideas generated in this process "won't necessarily be in *that* assessment, but probably by the time the next bloody assessment comes, people have gone away and thought about it and had a paper published on it. Then it's fair game for putting in the assessment."[113] Many participants agree that being part of an assessment is scientifically stimulating and can influence one's own future research in productive ways.

Bob Watson stressed both the social and the intellectual benefits that come from meeting experts in other fields whom one might otherwise not know, particularly since these are "the best" people in the world—and therefore people one really wants (and perhaps needs) to know:

> Why do people take part in these? Well, first, they meet their other colleagues. If you're a modeler, you meet all the other modelers. But even more important, you interact even closer with the experimentalists, etc., so you're meeting with the best people in the world right across the issues. . . . Especially scientists from developing countries have a real opportunity to meet the best from the US or the UK. The reason people keep doing this is threefold. One, they want to get

> their science reflected in this. Two, they find it exciting working with other scientists and learning things they hadn't learned on issues they might not spend much time on. And three, their science is influencing policy.[114]

Watson's third reason why scientists participate in assessments—"their science is influencing policy"—was not mentioned by many other interviewees. One might think that scientists would welcome the opportunity to make a difference in the world by connecting their scholarly work to matters of public policy, yet few scientists mentioned and none emphasized it. Rather, they emphasized the ways in which participation helped them qua scientists. Perhaps it is simply that as scientists they are most motivated by scientific activities; problems of public policy are not their priority. (An assessment involving economists might be a different matter.) Or perhaps this is related to the desideratum of perceived objectivity. A scientist who actively wants to influence public policy may be seen as not acting like a scientist. One who actively tries to influence policy—as Rowland did—might be considered biased. A scientist who wanted to participate in future assessments might consider it best not to express the desire to influence policy—or at least not to express it too strongly.

Disagreement

One question that emerged early in the ozone assessments was how to handle disagreement. Watson recalls:

> In the late 1970s there was a whole series of ozone assessments [and] certain politicians asked what were the differences between them rather than asking what the similarities were between them. A guy called Guy Brasseur[115] was once asked to write a short paper on the differences between them, which I thought was singularly unhelpful.[116]

Watson pushed for a single, comprehensive international assessment in part to counter any possible claim that any particular assessment was biased toward the policy position of its country of origin, and he wanted to focus on the areas of agreement, not disagreement, for fear that people or governments would use those differences as an excuse for inaction.[117]

Watson was certainly right that disagreement could be used as an argument against action, insofar as industry groups, particularly in the United Kingdom, were at that time pushing hard against aggressive action.[118] Understanding the areas of scientific agreement is crucial for evaluating where scientific knowledge provides a stable basis for decision making. But articulating areas of disagreement is important, too, since these may reveal areas where scientific knowledge is still in flux, as at that time it was.[119] Moreover, one might suppose that an assessment that clearly laid out the varying views of scientists, without enforcing consensus but giving full and free expression to whatever diversity of opinions existed, would have been easier to defend against accusations of bias than a report that attempted to present a "unified front."[120]

Because they interpreted the problem of bias primarily as a question of nationalism and the impact of the pluralism of national policy preferences, Watson and his colleagues opted for a single, international voice. Rightly or wrongly, their perspective on the political dimensions of the problem—that a single voice would be more defensible than a multiplicity of voices and therefore more effective in moving policy—structured the manner in which they designed their assessment.

One strategy for achieving a single voice was through the use of executive summaries. The first two major international ozone assessments, *The Stratosphere 1981* and the Blue Books (*Atmospheric Ozone 1985*), did not have executive summaries, an omission that Watson described as "a monstrous mistake."[121] The *Report of the International Ozone Trends Panel* had an executive summary, but it was published almost two years before the report itself and was not included in the full report once it finally became available. Since then, all of the international ozone assessments have included executive summaries as part of the assessment document. Watson described these as "absolutely critical," believing that the executive summary is the only part of an assessment likely to be read by policy makers.[122] Some scientists worried that policy makers don't even do this much but rely instead on underlings to condense the summary still further.[123] To these scientists, it is essential that the summary present the critical take-home conclusions as concisely and clearly as possible.

Watson and Albritton adopted a particular style for the ozone assessment executive summaries, designed to ensure that if they only had a couple of minutes to speak to a policy maker, they would be able to communicate the assessment's major conclusions and implications:

> We would just take all the bolds [i.e., bold points from each chapter of the assessment] put them together, print them out and say "Does this give you the ten most important messages?" In other words, if you had a minister or a major person from the business sector, and you were going up in the elevator [with

> him or her], all you would have time for is these bolds. Do they make a consistent story? Have we grabbed all the key information, the key messages? Have we forgotten any? . . . [In short] we don't waste any time. . . . [The executive summary] is absolutely focused on the major conclusions.[124]

Uncertainty

While scientists involved in ozone assessments grappled with boundary issues—whom to include, whom to exclude, what to include, what to exclude—they also grappled with the question of uncertainty: how to evaluate it and how to express it. Uncertainty is at once an internal and an external problem for scientists. Internally, understanding uncertainty—particularly in the form of error estimates—is a standard part of scientific practice and often considered part of scientific ethics. Scientific honesty includes being forthright about the limits of your knowledge, the uncertainties of your estimates, and the measurable errors in it. Uncertainty also has an external dimension, made manifest by the assessment process: if decisions are being structured in part by the knowledge expressed by the assessment, policy makers need to know how secure that knowledge is. If those decisions are designed to avoid future adverse effects, then uncertainty is part and parcel of the decision-making process as well as the expression of the knowledge on which it is (at least in part) based.

Scientists understand uncertainty primarily in terms of error bars—including the familiar statistical measures of means and standard deviations—but uncertainty may also take forms that are not readily measured, including areas in which relevant knowledge does not exist or is incomplete, areas in which available data do not permit conventional probabilistic statistical analysis, and areas in which scientists—for whatever reason—do not agree.

The international ozone assessments have never had a formalized procedure for handling uncertainty. Scientists working in recent years on climate change—where the handling of uncertainty has been formalized—have suggested that ozone was a simpler problem, less riven by uncertainty than climate change. For example, Brian Toon, an atmospheric physicist known for his work on nuclear winter who was also involved with several ozone assessments and the 1987 Airborne Antarctic Ozone Experiment, has argued that in comparison to climate change assessment, ozone was easy. Climate, he noted, to a large extent involves trying to predict things that have never been observed (such as ice sheet collapse); ozone, so he said, is more concrete, immediate, and certain. After the Antarctic ozone hole had been discovered, "the uncertainty of the ozone hole problem was more a matter of what did happen," so "one of them [climate change] is a prediction and the other one [ozone depletion] is something you observe."[125]

But the qualifier "after the Antarctic ozone hole had been discovered" is significant. In the early years of the assessments, ozone depletion was a prediction for which there was no observable evidence. Contrast this with the empirical evidence for climate change, which was already beginning to emerge when the IPCC was created. Indeed, when severe ozone depletion was observed, many scientists at first disbelieved their own data. Once they did come to believe it, they went scrambling back to their drawing boards to figure out how it had happened, since the existing models had not predicted it and could not account for it.

University of Cambridge atmospheric scientist Michael McIntyre expressed a view similar to Toon's. He noted, in a discussion of uncertainty treatment in assessments,

> The ozone problem is a much easier case than the climate problem. [With] the ozone problem, you can say quite a lot of things without bothering with numbers because they're practically certain. . . . You can say some things on the climate problem, but you can't say very much in terms of predicted scenarios.[126]

This was not how scientists viewed it at the time, however. The actual history of ozone assessments shows that uncertainty was a major issue in evaluating the threat of ozone depletion. Ozone may seem easy in retrospect because it is viewed today as an issue that has been largely resolved, but it did not seem simple to the scientists who worked on it in the 1970s and 1980s. Nor did the science seem certain to the chemical industry, which for a time pushed back strongly, or to the media, which reported the issue as one that was highly contested. The very fact of multiple assessments before the adoption of the Montreal Protocol, and the requirement of ongoing assessments to support it after its adoption, stand as testimony to the uncertainties. Multiple assessments were deemed necessary to cope with the complex, changing state of scientific knowledge.

As science policy scholar Henry Lambright has observed, the Montreal Protocol's provision for future revisions according to periodic assessments was an adaptive management framework, intended as a means to handle uncertainty:

> The MP [Montreal Protocol] included measures for subsequent review and amendment based on further scientific research. The policy makers understood

> that they were making policy and setting rules to contain CFCs under conditions of uncertainty. The delegates virtually invited the scientific community to reduce uncertainties in order to guide policy.[127]

Similarly, in a 1997 presentation celebrating the 10th anniversary of the Montreal Protocol, Dan Albritton referred to the iterative assessment process as "a crucial part of the structure of the Montreal Protocol," in which "the best opinion is given at the moment and the best decisions are made and revisited some years later."[128]

Metrics

The development of new metrics for conceptualizing ozone depletion has been an important part of the development of the science, a strategy for taming uncertainty, and a tool for helping policy makers glimpse the future.

As we saw earlier in this chapter, by 1988 scientists were confident that the Antarctic ozone hole was largely caused by heterogeneous reactions occurring on the surfaces of polar stratospheric cloud particles. But measurement difficulties continued to inhibit experiments aimed at characterizing those reactions, making it difficult to incorporate them into models used to predict future ozone depletion. (This remains a challenge today.) Scientists addressed this difficulty by developing a series of metrics for ozone depletion that in effect served as surrogates for predictive photochemical/dynamical models (see table 3.3).

These metrics were originally developed to give scientists and policy makers a way to gauge the relative risk associated with various ODSs (relative to each other in terms of their chemical properties and also relative to their respective emissions profiles). However, after the ozone hole discovery, they also served as a useful device for facilitating predictions of future depletion levels while avoiding the complexities associated with incorporating the heterogeneous reactions into models.[129]

The first ozone metric developed was ozone depletion potential (ODP), proposed by atmospheric modeler Don Wuebbles in 1983 and first appearing in an ozone assessment in 1989.[130] The ODP incorporated information about the composition and atmospheric lifetime of a given ozone-depleting substance; results were presented relative to the action of an equal mass of CFC-11 (i.e., CFC-11 was arbitrarily assigned an ODP of 1). With the recognition of the importance of additional ozone-depleting compounds (particularly those containing bromine), as well as the role of heterogeneous

TABLE 3.3. Metrics for estimating ozone depletion.

Ozone Metric	*Calculated By*	*Interpretation of Results*
Ozone depletion potential (ODP)	How much ozone depletion a fixed quantity of a given substance will cause, over its entire atmospheric lifetime, relative to the same amount of CFC-11	Need to combine calculated ODP with actual quantity released to estimate actual ozone depletion from this chemical
Chlorine loading potential (CLP) or bromine loading potential	How much chlorine (or bromine) a fixed quantity of a given substance will deliver from the troposphere to the stratosphere relative to the amount of chlorine delivered by the same quantity of CFC-11 over their respective lifetimes	Need to combine CLPs for all ODSs with actual quantities of all ODSs released to figure out actual ozone depletion
Equivalent effective stratospheric chlorine (EESC)	Estimates the total number of chlorine (and bromine) atoms delivered to the stratosphere, over a given time period	EESC is proportional to the amount of chlorine (and bromine) available in the stratosphere from all ODSs in a given period; allows easy calculation of indirect (cooling) contribution to the global warming potential of these ODSs

chemistry, ODPs were supplemented by another proxy metric: the chlorine loading potential, or CLP.

The CLP, which first appeared in a July 1988 report of the US EPA and was also used in the 1989 ozone assessment, represented how much chlorine (or equivalent ODS such as bromine) from a given halocarbon would be delivered to the stratosphere.[131] By measuring an intermediate step in the ozone depletion process, the CLP obviated the need to work out all the reactions through to the end point of the ozone-destroying step. Parson has credited this move as highly effective from a policy perspective: "Stopping the analysis [of projected ozone depletion by a given substance] at the intermediate step of chlorine loading avoided all the complexity, uncertainty, and controversy associated with the actual ozone-loss processes."[132] In other words, it was a scientifically defensible and quantitative means for dealing with the severe uncertainties still hovering around the actual ozone-destroying reactions themselves.

In 1994, a third metric was introduced: the equivalent effective stratospheric chlorine (EESC). Resembling the concept of "CO_2 equivalent" now used in estimating the effect of diverse greenhouse gases in climate change, the EESC provided an estimate of the total amount of ODSs delivered to the stratosphere that was available to destroy ozone at a given time. In addition to serving as an easily interpreted measure of a particular chemical's contribution to halogen loading and ozone depletion in the stratosphere during a period of decreasing production of ODSs, the EESC had another purpose: it provided a means of incorporating into the global warming potential for a substance the indirect radiative (cooling) effect of ozone depletion caused by a chemical.[133] With this new metric, policy makers could establish the net contribution of a given ozone-depleting chemical to climate change—a concern that, by the early 1990s, had risen to prominence and which, at the behest of the governments involved, the assessments began to consider alongside ozone depletion.

Since 1998, most of the assessments' graphical predictions of stratospheric chlorine levels under various emissions scenarios have been presented in terms of EESC. The EESC projections of ozone depletion agree very well with results from recent models, which embody a relatively complete understanding of ozone chemistry. In particular, the EESC-based projections are in good agreement with the full models with respect to the period when the ozone layer will return to "normal" levels over Antarctica. In essence, scientists combined the EESC with actual observations of the size of the ozone hole in recent years to project future ozone losses while avoiding the need for a detailed understanding of the chemistry.

CONCLUSION

Ozone assessment changed over time. It became large, international, and inclusive, and these characteristics helped to drive out an earlier model of nationally or institutionally based assessments. This shift occurred as some influential scientists came to believe that a single, inclusive international assessment would be more authoritative and responsive to the needs of policy makers than a multiplicity of smaller, exclusive ones.

As ozone assessment developed, scientists became increasingly self-conscious about the boundary between science and policy, promoting the idea of "policy-relevant but not policy-prescriptive" information, a distinction that lives on explicitly in the policies and practices of the IPCC.

Watson saw this distinction as critical. There are two reasons, he argued, for scientific assessments to avoid making policy prescriptions. First, if you do present policy recommendations, "you can always be viewed as [if] you're advocates and you've preselected the information [in the assessment] to support your policy position." Second, he noted that policy decisions are based on more than science:

> Scientific knowledge ... [is] only one input to a policy decision. There are other factors. There is the economy in the broad sense of the word, there's employment, there's distributional issues.... So for scientists to believe that they know what the answers are is a little bit—arrogant is too strong of a word—but it fails to recognize the other dimensions of a policy decision.[134]

Other scientists agreed. Acknowledging the limits of scientific expertise (and perhaps implicitly acknowledging the complex character of democratic decision making), Tony Cox stated that "policy options are not just scientific. They include all sorts of other things in them."[135]

Recognizing this, Watson and Albritton sought instead, in later ozone assessments, to present scenarios and state what the likely outcome of each one would be:

> The one golden rule of all of these ozone assessments and climate assessments and biodiversity assessments [i.e., all the assessments Watson has managed] is never to be prescriptive. My most favorite of all statements I've ever written in my life was one that Dan Albritton and I wrote at probably three o'clock in the morning, which of course we took back to all the other scientists.... It was basically on polar ozone, Antarctic ozone, and the statement went along the lines of, "Unless there is a 100% elimination" (or some words like this, or complete elimination) "of all long-lived chlorine- and bromine-containing compounds into the atmosphere, the Antarctic ozone hole will be with us forever." ... So it didn't tell governments what to do, but it told the consequences of not doing something. It was totally nonprescriptive, and yet it was really saying, "You've got to get rid of it [i.e., the anthropogenic chlorine and bromine compounds] if you want to get rid of the ozone hole." ... We've used a lot of "if-then." "If you want to achieve this, then these are the options." Well, you can do the reverse: "Without doing this, then...." The if-then approach is a very powerful approach because it's not prescriptive. If you do this, these are the consequences. If you want to achieve this, these are the options.... But the if-then is a very powerful approach. Telling governments what to do is not very useful. They tend to rebel against it.[136]

This view—that scientists do best by presenting options or scenarios rather than recommending policy—is now widespread, even dominant, in the scientific community.

However, different scientists have different understandings of what this distinction means in practice. Albritton stresses that the role of scientists is to present a small number of sharply distinct options: "We . . . learned that in talking to policy makers, rather than presenting fifty choices, present four choices, spread wide enough that there is no doubt that the science is robustly saying that one is different from the other."[137] Modeler Don Wuebbles (who developed several ozone metrics and participated in many of the assessments) argues that scientists have a role to play in evaluating options as well as presenting them:

> Why shouldn't we be involved with talking with policy makers about potential impacts and various options? I think in the assessments, we want to be very careful to not prescribe for them exactly what things they should do, but I think it is useful to tell them and ask those "What if?" questions. If you did this, the impact it would have on ozone, based on our best knowledge.[138]

Jonathan Shanklin, a British meteorologist and one of the discovers of the Antarctic ozone hole, has expressed the view that "what you have to do is give policy options and the consequences, and then the politicians can choose from a menu as to what they see as the least worst from their perspective." However, other participants in the seminar in which Shanklin expressed these views challenged his use of the phrase "policy options." Atmospheric scientist Michael McIntyre queried, "That's getting the two things tangled, isn't it? The scientists should say what the facts are as far as we can tell; the policy makers should look at policy options." As a friendly amendment, he proffered: "I think . . . in scientific assessments, [we should say] if you do this, then we think the range of possibilities is that. End of story."[139]

The development of ozone metrics was tuned to this project of providing policy makers with scenarios and options in the face of uncertainty. Metrics were originally designed to inform decisions on trade-offs between limiting production of one ODS versus another by how much and when. Assessments subsequently repurposed them to allow projection of future ozone depletion absent models that were adequate to the task.

But while scientists were developing metrics useful to policy makers, the natural world offered up a surprise: the discovery that matters were worse than scientists had imagined. Assessments inevitably leave things out, not just for political or social reasons, but also for scientific ones, and what gets put in and what gets left out scientifically is a major issue affecting the stability, robustness, and ultimately the reliability of the conclusions of the assessment. In scientific research, difficult problems are often set aside on grounds of intractability, with the hope or assumption that progress can meanwhile be made in other areas.

Simplifying assumptions are part of scientific research, as they were in this case.[140] In the history of ozone assessment, heterogeneous chemistry was omitted from serious consideration until the unexpected discovery of the Antarctic ozone hole forced scientists to reexamine their views. In this case, it was not the assessment itself but discoveries in the natural world that forced that reexamination. In other cases, the pressure on assessments to explain unexpected results surely contributed to a sense of the necessity—the urgency—to reexamine previous assumptions to determine what, in those assumptions, might have been incorrect. As we shall see in the next chapter, assessments are sometimes unable to reconcile such issues satisfactorily. The history of ozone assessment reminds us that assessments are not the same as—and not a substitute for—continued basic scientific research. Indeed, it underscores the imperative of continued direct engagement with the natural world.

OZONE TIMELINE

1974

- Stolarski and Cicerone paper shows a chlorine chain reaction in the stratosphere could destroy ozone
- Molina and Rowland paper shows chlorine from CFCs is reaching the stratosphere, discusses various catalytic cycles destroying ozone

1976

- First US NAS (panel and committee) ozone reports released
- First UK Department of the Environment report published

1977

- Ozone protection amendment added to the US Clean Air Act

1979

- Second US NAS ozone report released
- Second UK Department of the Environment report published

1985

- March: Vienna Convention for the Protection of the Ozone Layer adopted
- May: British scientists publish first paper identifying the Antarctic ozone hole

1986

- July: "Blue Books" assessment (*Atmospheric Ozone 1985*) published
- August–September: NOZE expedition to study Antarctic ozone hole

1987

- August–September: NOZE II and AAOE expeditions to study Antarctic ozone hole
- September 16: Montreal Protocol on Substances That Deplete the Ozone Layer agreed to
- NASA et al. release initial findings of Antarctic expeditions, showing convincingly that the ozone hole is caused by chlorine chemistry

1988

- March 15: Ozone Trends Panel Report executive summary released, demonstrating small but noticeable global ozone depletion (in addition to the Antarctic ozone hole)
- March 18: DuPont announces it will stop manufacturing CFCs within 10 years

1989

- January 1: Montreal Protocol enters into force; mandates ongoing ozone assessments at least every four years, beginning in 1990

1990

- *Scientific Assessment of Stratospheric Ozone: 1989* published, the first scientific assessment published under the Montreal Protocol

CHAPTER FOUR

Assessing the Ice: Sea Level Rise Predictions from the West Antarctic Ice Sheet, 1981–2007

INTRODUCTION

In May 2014, the *New York Times* reported, "A large section of the mighty West Antarctica ice sheet has begun falling apart and its continued melting now appears to be unstoppable."[1] Based on papers published that month in *Science* and *Geophysical Research Letters*, the article suggested that the retreat of the West Antarctic Ice Sheet (WAIS) had passed its tipping point. Since the late 1970s, scientists had feared the loss of the ice sheet as a result of climate change; now that fear was being realized. Seasoned WAIS geologist Richard Alley, who had long known that WAIS disintegration was possible, was quoted as saying, "It shook me a little bit."

The *New York Times* story privileged two points in the history of studying the WAIS: a 1978 prediction of its collapse and the present-day assertion that collapse was now inevitable. This might have seemed to suggest that scientists had been expecting this news—that their early expectations were now being fulfilled. But this was far from the case. For more than three decades, a substantial number of glaciologists had interpreted the available evidence as indicating that the WAIS was relatively stable and would remain so for centuries or even millennia. They did not see its possible collapse as a threat that policy makers needed to consider seriously. Yet at the same time, a significant number of glaciologists worried that collapse might occur far sooner and that the threat was something that policy makers did need to heed. The scientific finding that WAIS disintegration was now unstoppable—and might unfold within the foreseeable future—could therefore be understood either as an overall reversal of scientific opinion in response to new evidence or as a perhaps temporary triumph of one unproven view over another. Either way it raises the question: Why did many glaciologists for so long judge the WAIS to be stable until they decided that it wasn't?

Historically, the WAIS has puzzled glaciologists trying to understand its behavior and future. As a marine ice sheet in contact with warming ocean waters, it is inherently vulnerable to long-run disintegration. Scientists already understood in the 1960s that if it were to disintegrate, it would raise global sea level by three to six meters.[2] However, as far as glaciologists were able to judge at that time, WAIS did not appear to be disintegrating; if anything, it was growing. Any possible disintegration seemed to be a very distant threat, especially remote when measured against political time frames. However, by the early 2000s, as the *Fourth Assessment Report* (*AR4*) of the Intergovernmental Panel on Climate Change (IPCC) was under way, it had become apparent that the WAIS was less stable than previously thought; in fact, some areas of the ice sheet already showed evidence of rapid change. These findings led scientists to reexamine previous sea level rise projections for the twenty-first century.[3] This major change in scientific understanding led to a comparable shift in the level of concern about rapid disintegration as a socially significant impact of global warming.

In the 1950s and 1960s, a few key US scientists dominated the discussion of anthropogenic climate change, and the future of earth's ice sheets was among the issues that they highlighted. In the 1970s and early 1980s, major differences of opinion emerged as to how climate change would affect the WAIS, in particular whether a rapid disintegration in the near term (i.e., less than two centuries) was possible. This motivated WAIS scientists to organize workshops to sort out the varying views and attempt to define a policy-relevant research agenda. As the science grew and became more diverse and more complex, changes detected in the cryosphere and in global sea level inspired interest and concern among a much larger group of scientists. These scientists and the early workshops they instigated shaped many of the questions still pursued by the research community and addressed by scientific assessments today.

This chapter traces the history of the WAIS assessment process, an intricate story of epistemic shifts in which the evolution of scientific knowledge was closely intertwined with its assessment. We follow its evolution from small-scale, informal workshops and single-authored reports to the elaborate, international, and formal assessments produced by the IPCC. We focus on US and European WAIS assessments up to the fourth IPCC report, comparing pre- and post-IPCC WAIS assessments and the shift in scientific characterizations of WAIS. In particular, we try to illuminate how scientists evaluated the potential threat of a rapid WAIS disintegration, why a majority of them for so long judged that threat to be modest, and how they have recently come to view the matter differently. Over time, WAIS assessments became subsumed by the IPCC assessment process as part of the larger project of evaluating and projecting the effects of climate change. Because computer models of future climate were unable to fully account for observable changes in WAIS, models came to be regarded as inadequate to project future changes in WAIS. But rather than fully exploring alternatives to mechanistic modeling, IPCC's Fourth Assessment refrained from estimating WAIS's twenty-first-century contribution to sea level rise. As often occurs in areas of unsettled knowledge, assessment authors seemed anchored to "erring on the side of least drama."[4]

EARLY OBSERVATIONS ON THE WEST ANTARCTIC ICE SHEET

The International Geophysical Year (IGY) of 1957–1958 focused scientific attention on global data gathering in far-flung locales, including Antarctica. Under the aegis of the IGY, glaciologists began multiyear research on the characteristics and behavior of WAIS. Scientists participating in the IGY also put in place a system of stations around the world, including in Antarctica, that continue today as part of the Long Term Ecological Research network, which compiles significant data on the state of the planet's various ecological zones.[5]

For American scientists, the formation and formalization of National Antarctic Programs around the same time streamlined logistics and research funding, a welcome change from the sporadic and mostly territory-oriented expeditions of the previous era. The growth of the US National Science Foundation (NSF), which took on a leadership role in Antarctic research in the 1960s, further strengthened American Antarctic research.

The United States established several small, temporary bases during the IGY but then focused primarily on building a large base that could handle continental-scale logistics. Near the location where British naval captain Robert F. Scott had begun his ill-fated attempt to reach the South Pole, the United States established McMurdo Station. The scale of infrastructure at McMurdo, which grew considerably over the course of the IGY, simplified the always costly and difficult logistics of managing deep-field scientific research—work conducted in remote areas far from stations. (Today, McMurdo boasts a summer population of 1,100, with multiple bars, a church,

ATMs, a January music festival called Icestock, state-of-the-art laboratory facilities, several landing options for aircraft, and other amenities.)

US scientists also established Palmer Station, a small base on the Antarctic Peninsula, the most northerly and warmest part of the continent. The relatively diverse populations of plants and animals attracted biologists and ecologists to the peninsula; the station also supported some deep-field research. A third station at the South Pole was developed primarily for atmospheric and astronomical research. Having a base at the South Pole, which every Antarctic territorial claim save one included, also carried symbolic significance during the Cold War and beyond.[6]

One of the United States' key deep-field research programs was the study of the Ross Sea sector of the West Antarctic Ice Sheet. It spanned several decades, included various research teams from diverse universities, and consumed a significant portion of NSF's Antarctic research funding. The principal investigator of the early US-funded WAIS projects was Charlie Bentley of the University of Wisconsin. Following the IGY, Bentley spent the 1970s working on the Ross Ice Shelf Geophysical and Glaciology Survey and the 1980s with the Siple Coast Project. He also chaired the National Research Council (NRC)'s Polar Research Board, and by the end of his career had become the elder statesman of WAIS research.[7]

The first systematic, US-funded exploration of the ice sheet was a series of seismic traverses.[8] Considered the key US Antarctic project under the IGY, these traverses, undertaken in Sno-Cats, sought to measure the thickness of the ice sheet and, indirectly, the topography of the continent underneath (see figure 4.1). Bentley led several of them, stopping every three miles to collect measurements.[9]

But scientists did not just want to know how thick the ice sheet was; they wanted to understand how the ice flows in a dynamic landscape. A large team led by Barclay Kamb and Hermann Engelhardt from the California Institute of Technology (Caltech) conducted fieldwork immediately after Bentley's pioneering traverses and focused on ice streams—regions of the sheet that seemingly flowed toward the ocean between areas of relatively static ice, analogous to rivers flowing through a hilly landscape. This was very challenging work. The only other glaciological team on the ice around the same time was that of Ian Whillans, who was surveying Ice Stream B, later to be renamed Whillans Ice Stream.[10] In an interview, Engelhardt noted the difficulties of studying a location about which scientists knew so little, in terrain so treacherous that their research team lost an airplane in a crevasse.[11] Kamb had to ship his heaviest drilling equipment

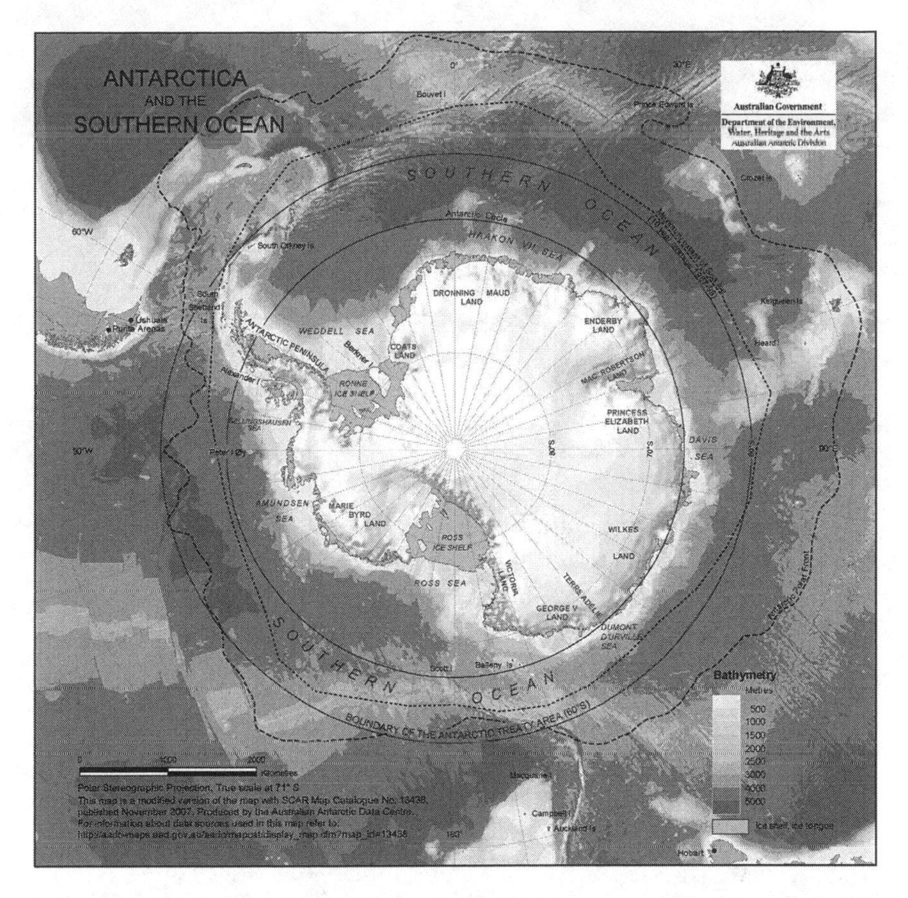

FIGURE 4.1 Antarctica and the Southern Ocean. Map ©Commonwealth of Australia, by the Australian Antarctic Division, Australian Antarctic Data Centre.

down to the ice by boat a year in advance because it was too heavy to be flown down.

Besides the incredibly complicated logistics, the scientists confronted the challenge that the landscape they surveyed literally shifted as they attempted to track it. The basic surveying work, which remains incomplete (though much improved), was conducted during Engelhardt's tenure on the ice; aerial surveys helped the Caltech team determine what parts of the Ross Ice Shelf were most active. The Kamb/Engelhardt Caltech team undertook to drill into a slowly moving ice stream now named for Kamb. Ice drilling was itself an emerging technology, and the group worked in their Pasadena lab to construct hot-water drills that could descend to the bottom of the stream, carrying equipment to photograph and collect samples along the way. Through this project, they were able to map the dynamic, fluid terrain underneath the ice, which included areas of flowing water, points where the ice was lodged against rocks or ridges, and areas that flowed slowly. Kamb and Engelhardt were also able to characterize some of the local sub-ice behavior of the West Antarctic Ice Sheet and the ground and water it rests on.

Glaciologist John Mercer from Ohio State University was known in the Antarctic scientific community as a bit of an oddball; stories trickled back from Antarctica about him conducting research in the nude.[12] In a conference paper in 1968, Mercer was the first to link atmospheric warming with the potential rapid disintegration of WAIS. The conference paper received very little attention; a decade later, he repeated his message in a paper with a dramatic title: "West Antarctic Ice Sheet and CO_2 Greenhouse Effect: A Threat of Disaster."[13] The use of the term "disaster" in the title was uncharacteristic of scientific writing, and it may have unsettled some of his colleagues, but the paper anchored virtually all subsequent scientific reviews—as well as disaster scenarios—of the subject. Mercer's speculation was based on evidence that global sea level was about six meters higher during a warm period 125,000 years ago, and lake sediments suggesting that WAIS had temporarily disappeared sometime in the past million years or so.

The argument gained support from Johannes Weertman's theory that WAIS and other marine ice sheets that are bounded by sea water and based on bedrock below sea level are potentially unstable. Their "grounding lines," demarking the boundary where land-based or "grounded" ice slips into the sea forming floating ice shelves and contributing to sea level rise, are susceptible to rapid retreat when the abutting ocean warms. Although ice shelves float, they are often stuck against surrounding cliffs or rocky rises in the sea floor, restricting their forward motion and hindering outflow of the grounded ice behind them. As ocean water warms, ice shelves melt more quickly, reducing this buttressing and allowing rapid disgorgement of grounded ice into the sea. This process could end with the entirety of WAIS being transferred from land to ocean.

EARLY US WAIS ASSESSMENTS

The Orono Conference: 1980

Mercer's 1978 *Nature* paper inspired the first consolidated, multidisciplinary overview of the state of WAIS science, held at the University of Maine,

Orono, in 1980. The Orono conference was jointly sponsored by the American Association for the Advancement of Science (AAAS) and the US Department of Energy (DOE) and anchored by University of Maine professors Terry Hughes and Jim Fastook.[14] A 558-page verbatim transcription of the formal proceedings provides a written record of both the presentations and the question-and-answer sessions that followed each presentation and concluded the workshop.[15]

Charlie Bentley organized the presenters. The distinguished oceanographer Roger Revelle, although not in attendance, wrote the charge letter outlining the intended product of the conference: a research proposal. Roger Revelle was an early player in American climate science.[16] He was instrumental in securing funding for Charles David Keeling's atmospheric monitoring project during the IGY, which led to the famous Keeling Curve (the graph that clearly depicts an anthropogenic increase in atmospheric carbon dioxide), and he continued to promote scientific and policy interest in climate change throughout his career.[17]

In the charge letter, Revelle focused on the need for the conference to articulate a research agenda:

> The report of this meeting should describe each research issue as explicitly as possible, what are the questions that need to be answered, what is the most productive way of approaching the task, what resources will be needed.[18]

The list of meeting presenters reads as a veritable Who's Who of West Antarctic Ice Sheet research: John Mercer, Terry Hughes, Jim Fastook, Bill Budd, Charles Raymond, Ian Whillans, Craig Lingle, Bentley, and many other glaciologists whose contributions to WAIS science continue to be influential today.

Welcoming remarks were given by Dr. David Burns from AAAS. At the time, AAAS and the DOE were putting together a research program to guide US climate research—and, in particular, establishing the National Climate Program and participating in the World Meteorological Organization (WMO)'s World Climate Program. Reinforcing Revelle, Burns reminded participants that the papers that emerged from the meeting should not be "scientific papers in the usual sense of presenting new information or results of experiments or data" but should summarize what was known and outline what needed to be known. He explained:

> There are two primary audiences for the product. . . . One is yourselves and your colleagues in the field of glaciology and Antarctic research and the paper or the research plan would be an indication to them of—of research that [it] is

> felt would be useful and would be a guide in effect to writing proposals. The other audience for the product of this meeting would be the bureaucrats and the administrators who will be receiving such proposals and will use your deliberations and the written result of it as a guide to funding research.[19]

Burns addressed the policy significance of future sea level rise, discussing an article by climate modelers Stephen Schneider and Robert Chen in the context of a recent congressional hearing:

> I can assure you that the possibility of a rise in sea levels and the loss of productive land are taken quite seriously. I don't know whether they [Schneider and Chen] were grabbing for headlines or not, but they were asking questions. We are now doing things about energy we are producing, we are making decisions about shale, we are making decisions about fossil fuel. These decisions will have results and if the hypothesis of carbon dioxide warming and climatic change is correct, the decisions we are making today about energy may have consequences that would be costly or unpleasant for human activities. So, I think that there is an urgent need to consider what the mechanisms and the likelihood of this occurring [are], and your deliberations will tell us how we can go about finding out what we need to know.[20]

It was not unusual for a government agency to sponsor a research conference to summarize the state of the science; what made this conference an assessment was the explicit recognition that a policy response to the science might be required.

Bentley reiterated the idea that the meeting was different from the usual scientific meeting or workshop: "We're in serious danger, I think, of going on in a very enjoyable discussion of scientific results, which is not the purpose for which we are here."[21] Instead, he stressed, the participants should carefully follow Revelle's charge: "While it may be necessary to include some retrospective material, the paper should not emphasize what we know now, but rather what we need to find out and how we should go about it."[22]

In articulating a research agenda, the scientists would have to address their different views about how WAIS might contribute to sea level rise and how imminent a threat that might be. Bentley's published research suggested little likelihood of rapid WAIS disintegration in the coming centuries, but Mercer, Revelle, and Hughes had offered data and scenarios that could put disintegration into play much sooner, within the next 100–200 years. Perhaps for this reason, Bentley kicked off the formal presentations by introducing Mercer with a small joke: he facetiously gave Mercer's speaking slot the title "The Threat." Mercer responded to Bentley in kind: "Okay, Charlie, yes, I had to expand that title a bit, I didn't see it until a few days ago, 'The threat reconsidered in the light of new climatic events during the last interglacial.'"[23]

Terry Hughes discussed CLIMAP (Climate Long-range Investigation, Mapping, and Prediction), an ambitious effort to produce a map of climate conditions during the Last Glacial Maximum. CLIMAP researchers had reconstructed prehistoric episodes of disintegration and retreat of WAIS toward the center of the continent and modeled potential future retreat. Hughes's presentation generated a long discussion about the equations he used and the specific ice sheet locations he described; one unidentified speaker suggested that the evidence Hughes had used to describe future disintegration actually hinted more clearly at near-future WAIS stability.[24]

Jim Fastook's presentation on the CLIMAP reconstruction, particularly the retreat that had occurred in the Pine Island Bay area, also generated significant interest. Bill Budd pressed for alternative strategies for depicting dynamic retreat.[25] The Pine Island Bay glaciers (which continue to be one of the most dynamic regions of WAIS today) featured prominently in discussions about possible disintegration scenarios explored by the CLIMAP team. Bentley pressed for an actual time scale for WAIS disintegration suggested by the CLIMAP model of the ice sheet's behavior. Fastook responded that a fast disintegration would take about 1,000 years, while a slower one might take place over 18,000 years.[26] Fastook also made clear the limitations of the CLIMAP model, focusing particularly on the model's inability to link the calving rate (the rate at which icebergs break off the edge of the ice sheet as it disintegrates) with the overall rate of ice retreat:

> At this point I don't have any way to verify [the model]. That's one of the hopes of tying the calving rate in some way to the ice velocity across the grounding line. I'd be happy with an empirical relationship between the calving rate and ice velocity, something—just something that I could use to take the parameter out of it, so to speak, and allow it to vary through the course of retreat.[27]

Fastook shared his findings and projections but acknowledged them as imperfect and puzzling—making clear that he saw this work as a step in the right direction but a partial one at best. Other participants were similarly open about the limitations of their work.

After lunch, the tone of the conference shifted markedly. Dave Drewry of the Scott Polar Research Institute opened his talk by saying, "Maybe I should start off with three very simple words," which were "I don't agree."[28]

His findings in the Ross Sea area contrasted strongly with the CLIMAP results. In Drewry's description, ice sheets were not configured so as to be vulnerable to warm ocean water, as suggested by Weertman, and this informed his opinion that WAIS was stable. After the presentation, the discussion of what WAIS grounding lines could look like (in reality as well as in models) and whether a true grounding line could even exist took up more time than had the presentation. This conversation eventually turned to the observational, geological research that would improve the modelers' work. Drewry summed up these research needs in the pithy incantation "We need more core, we need more core."[29] The modelers had a long list of locations about which they wanted to know more. The discussion continued with the presentations of Bill Budd, Uwe Radok, and Charlie Raymond, who complicated the CLIMAP model and offered observations to support, foil, and challenge the predictive work that WAIS modelers had just begun to perform.

The day closed with Burns returning to the policy implications of the work. Should WAIS concern policy makers? It all depended on the time scale:

> If you're quite certain that the time scale of disintegration of the Antarctic ice and the possible rise in sea level is a very, very long time scale, thousands of years, then I think that the policy people are going to lose interest quite rapidly. But if there is still quite a good possibility that it's much shorter than that, then I believe that the research would be supported, because ultimately the research is very simply to determine what our fossil fuel policies will be.[30]

On day two the scientific presentations continued, with each followed by lively debate. For example, following Ian Whillan's presentation, he and Dominique Raynaud disagreed on a fundamental question: Was the currently observed thinning of parts of WAIS a very short-term phenomenon or a longer-term process with a historical component, such as a past sea level change with effects only now reaching the interior?[31] Whillans and Mercer debated the length of time needed for WAIS to respond to sea level changes.[32] Bill Budd and others pointed out errors in Whillan's model, and so on. The workshop participants collegially and meticulously scrutinized the shortcomings of one another's work as well as their own. Budd, for example, noted that Bentley's and Whillan's findings conflicted, and Bentley responded by saying, "I'm looking at the whole area and he's talking about one flow line," suggesting that Whillan's smaller scale caused the contradictory findings.[33] However, later in the day, glaciologist Kenneth Jezek contradicted Bentley's assessment of a positive mass balance for WAIS (WAIS gaining ice) with new observations of a negative mass balance (WAIS losing ice) in the Ross Sea sector.[34] In these cases, WAIS glaciologists found themselves grappling with primitive models of a feature that did not behave uniformly in all locations or at all scales, and accumulation data (e.g., snowfall amounts) that sometimes had not been updated since the IGY, as well as the question of how to include uncertainty calculations in their predictions.

The postpresentation discussions circled back to the goals of the meeting—that is, to figure out what research trajectories were needed to understand WAIS more completely and predict its future behavior. Bentley divided the participants into smaller groups to hash this out. Then the workshop participants convened one last time on the morning of April 30, 1980. After a final set of research presentations, they laid out plans for the morning's small group discussions. Before they split up, British glaciologist Charles Swithinbank noted that some of the calls for satellite remote sensing might "price ourselves right out of the business" of WAIS research, particularly among scientists who came from less wealthy nations.[35] He suggested that they brainstorm some less expensive options.

Ironically, the workshop proceedings did not include a list of research priorities, only documentation of the presentations and group organization that facilitated its creation. However, an edited list of research priorities agreed to at Orono showed up in Roger Revelle's papers at the Scripps Institution of Oceanography (SIO) archives as an appendix to an SIO WAIS workshop held two years later:

> VII. *PRIORITIES*
> Projects and parts of projects are divided into three groups. There is no priority ordering within groups.
>
> Top:
>
> Satellite altimetry
> Satellite monitoring of Antarctic sea ice distribution
> Amundsen Sea drainage basin study
> Interaction between ice shelves and ocean
> Grounding line studies (including ice rumples)
> Drilling on ice rises and ice domes
> Deep drilling in East Antarctica
> World-wide sea level and ocean volume changes—last interglacial
> Ocean, sea ice, and precipitation modeling using global circulation

models
Numerical modeling of ice sheet response
Technique for surface mass balance from satellites
Radar profiling of ice shelves for bottom balance rate
Radar mass balance on ice shelves
Oceanographic monitoring along barrier
Instrumental emplacement deep in ice shelf for bottom balance rates
Marine coring for evidence from last interglacial and Holocene retreat record
Tide gauge studies

Second:

Ross Sea drainage basin study
Satellite studies of ice margins
Ice thickness mapping
Development of laser ranging capabilities
Marine coring for evidence of extent of last major ice advance

Third:

Weddell Sea drainage basin study
Bellingshausen Sea drainage basin study
Study of existing cores
Shallow coring
Antarctic Holocene sea level changes
Monitoring temperature change
Monitoring accumulation rates on stake networks[36]

Was this what Burns wanted, given his focus on providing a time frame for policy makers? Or did the list enumerate what scientists would have done anyway? The length of the list suggests that scientists were not so much thinking about answering critical policy-relevant questions as they were thinking about all the scientific components needed to truly understand WAIS.

One key difference between Orono and later assessments and workshops was the openness of the dialog during the meeting. That a record was made for public consumption is itself noteworthy. There were certain presentations and conversations where the reporter noted that the group had gone off the record. Still, the text recorded a great deal and provides a close look at how professional glaciologists struggled with this massively difficult problem. The participants were open and collegial and joked with one another but not without being critical or questioning, and well aware of the gaps, partialities, and uncertainties in their research.

Revelle, Bentley, and Changing Climate: 1983

In his papers, Revelle kept a 1981 special publication called "Carbon Dioxide Proliferation: Will the Ice Caps Melt?" from the Institute of Electrical and Electronics Engineers Power Engineering Society. The meeting that produced this publication included participants from both government and academia. It featured meteorologist Dr. Joseph O. Fletcher, who served as the meeting's expert on CO_2 warming in the polar regions; he had spent much of his research career in the Arctic, had served as the head of the NSF's Office of Polar Programs from 1971 to 1974, and at the time of the workshop was the acting director of the Environmental Research Laboratories at the National Oceanic and Atmospheric Administration (NOAA).

Fletcher had a large task at this workshop—discussing all of the ice in the polar regions, not just WAIS. When it came to WAIS, Fletcher's message was that the ice sheet would not pose a problem in a warming world. He claimed that if the ice sheet disintegrated, it would simply refreeze when it hit the water.[37] Fletcher also discussed the probability of surges in the rate of WAIS disintegration, offering two possible scenarios. First, he discussed the warming air temperatures and restated his claim about refreezing. The second scenario envisaged a warming ocean interacting with the ice sheet's edge. Here he claimed that "it is not likely that a warmer sea would occur, even with the warming by CO_2,"[38] because the severe cooling effect of the winds off the Antarctic land mass, combined with the hypercooled Antarctic Bottom Water, would keep the ocean from warming even if the atmosphere warmed as much as 8°–10°C over the Antarctic.[39] This was an extremely conservative, perhaps even complacent view of the possibility that WAIS would respond to warming, but he argued that it was warranted given the doubtful reliability of model results. "I hate to be such a skeptic," he claimed, "but I do think that although the modeling experiments are powerful tools, we have to interpret them with a great deal of caution."[40]

In 1982, Revelle reviewed the state of the science for *Scientific American* in a paper entitled "Carbon Dioxide and World Climate." Revelle said that the answer to the question of whether anthropogenic climate change was under way was "probably yes."[41] Bentley, who focused only on ice, saw this issue as far off in the future, but Revelle—who was very close to the modeling and empirical results and thus concluded that anthropogenic warming had likely already begun—saw the potential impacts on WAIS as a nearer threat.

In 1982 Revelle chaired an informal workshop at SIO, "CO_2, the West Antarctic Ice Sheet, and Global Sea Level." Revelle was now emeritus director of SIO; the directorship of SIO was in the hands of physicist William Nierenberg. Nierenberg was also chair of the National Academy of Sciences Carbon Dioxide Assessment Committee (CDAC), which the US government had charged with putting together the climate science assessment *Changing Climate,* published in 1983. The results of the SIO workshop would inform Revelle's contribution to *Changing Climate,* a single-authored chapter entitled "Probable Future Changes in Sea Level Resulting from Increased Atmospheric Carbon Dioxide."[42]

Twelve people participated: two notetakers; one government liaison from the Department of Energy; Bill Nierenberg (representing CDAC); and eight expert scientists—Revelle, Bentley, Arthur Bloom (Cornell), Wallace Broecker (Lamont-Doherty Geological Observatory at Columbia University), Edward Epstein (NOAA), Robert Etkins (NOAA), Vivian Gornitz (Goddard Institute), and Richard Wetherald (Geophysical Fluid Dynamics Laboratory at Princeton). Bentley was the only link between the Orono and SIO events, but he strongly advocated for the research needs articulated in Maine.

The workshop was organized around the question of threat: "How rapidly could the present WAIS disintegrate?"[43] In the opening remarks (and after briefly describing CDAC's mission), Revelle plunged into a discussion of his definition of rapid disintegration: "If [WAIS] were to disintegrate over several hundred years, we could no doubt adapt. But if significant rises in sea level took place in 100 years or less, there would be real problems."[44] Revelle also noted that "the WAIS could be monitored directly, but, this is not yet being done and there is no agreement yet on what are the most important measurements."[45] This introductory statement went directly to the question of what information was policy-critical and what was superfluous.

Wetherald was the first presenter, and he discussed the applicability of numerical climate models to WAIS. The models of the time showed an intact continental ice sheet at atmospheric CO_2 levels up to four times the current amount. But were they right? He noted that the CLIMAP "model was poor for Antarctica."[46] The oceans did not circulate, the polar surface air temperatures were too warm, the clouds did not change in response to changing CO_2 levels, and sea ice and ice shelves were not differentiated.

Bloom presented on paleoclimate and the history of global sea level. A hundred and twenty thousand years ago global sea level was six meters higher, making it likely that WAIS had disappeared during that time.[47] He discussed the complexity of determining prehistoric sea level, in that scientists had to consider tectonic uplift in relationship to changing sea level. The workshop reporter charmingly noted that "the discussion on proving 6 m higher sea levels about 125,000 years BP survived the break for lunch." The discussion apparently hinged on a debate between Broecker and Bloom, with Broecker arguing that the dates of this period of high sea level, called the Eemian,[48] were unproven; Bloom disagreed.[49]

Etkins and then Epstein presented on contemporary sea level rise. Etkins discussed differentiating the polar ice contributions to sea level rise from the thermal expansion component. Epstein described how he extrapolated warming-related sea level rise in contrast to that resulting from changes in the earth's rotational inertia (causing the ocean to bulge in places) and suggested that satellite altimetry might be the only reliable check on his WAIS data. The presentation spurred a conversation about whether researchers were gathering reliable global temperature data at all, since some findings, such as James Hansen's, seemed to be at odds with the data used by Epstein and others.

Gornitz—the only woman present—gave a presentation on tide gauge data based on an article she had recently published in *Science*.[50] Gornitz's findings suggested a recent sea level rise of 1 mm/year, less than Ken Emery's 3 mm/year. This ended the formal presentations for the first day of the workshop.

Bentley kicked off day two with a primer on dynamics focused on the logistics of how, when, how fast, and where a WAIS disintegration might occur. He described the ice sheet as in balance: "The sum of the thickening-rate at the margin and the bottom melt is zero."[51] He also insisted that while WAIS was unstable, there was "no way" that it could disintegrate in the next century.[52] Broecker, who tended toward rapid disintegration, asked Bentley if it looked like WAIS was "setting up for a surge [into the sea]"; Bentley did not think so. He also noted that by looking at the flow lines in the ice, one could conclude that there had not been a fast surge in 1,500 years.[53] Nierenberg, who tended toward thinking any threat was very remote, suggested that Bentley's data allowed for a positive ice thickness gain of up to 20–30 mm/year due to accumulation.

Broecker also asked about the mechanics of a warming Southern Ocean sending a warming signal up the Pine Island Glaciers, potentially triggering a very rapid disintegration once the ice became unhinged from its island pinning points. Bentley revisited this after lunch, to the point of belaboring the answer that "with a rise in temperature, chances of eventual disappearance of the WAIS are good, but that chances of its disappearance in a couple of hundred years are *not* good."[54] In his view, the 100-year interval for disintegration was impossible and the 200-year interval nearly so.

In a set of final remarks, Revelle staked out the possible WAIS contributions to sea level rise:

- Ablation [ice melt, from the continent overall] might furnish ≤ 3 mm/yr
- Pine Island increased ice motion ≤ 3 mm/yr sea level equivalent
- Retreat of ice margin in Pine Island Bay resulting in widening front ≤ 3 mm/yr
- Increased calving in Pine Island Bay due to thinning ice from widening front and bottom melt ≤ 3 mm/yr
- Ronne-Filchner shelves increased velocity 10 km/yr over a 1000-mile front ≤ 15 mm/yr.[55]

The sum of these contributions was equivalent to 27 mm/year, or 5.4 m of sea level rise in 200 years—assuming that disintegration would start at full speed.[56] Summarizing, the reporter wrote: "The sense of the meeting was that it was most unlikely that disintegration of the WAIS could result in a sea level rise as great as six meters in 100 years, but that there is an upper limit possibility that, including *all* the shelves (Pine Island, Ross, and Filchner-Ronne) a three meter rise could occur in 200 to 250 years."[57] Though this statement seems out of line with Bentley's views, there is no record of assent or dissent from Bentley at this point.

The workshop summary ended with the sentence "Revelle then listed the needed research as given in the summary."[58] Bentley furnished this list, attached as an appendix: an edited version of the list already generated at Orono. In this way, the as-yet-unpublished results from Maine made it into the proceedings of the SIO WAIS workshop and would influence Revelle's chapter in *Changing Climate*.

However, the SIO workshop participants did not simply repeat the Orono research priorities; they recrafted this list, removing the prioritization but adding more specific detail about certain measurements and data needed, as well as the locations where they should be obtained:

1. Satellite observations of surface elevation of the WAIS with laser altimetry at most sensitive spots
   a. Inflection points of ice-streams
   b. Grounding lines of shelves

c. Above them, away from the ice streams
d. Snow accumulation rates from microwave measurements
e. Profiles in ice streams

2. Measuring bottom melt underneath ice shelf
3. Drilling through WAIS to determine oldest ice in Siple Dome, Crary Ice Rise, Roosevelt Island
4. Mass balance study of drainage basins
   a. Amundsen Sea, Pine Island, and Thwaites Glaciers
   b. Ross Sea
   c. Weddell Sea (FRG [Federal Republic of Germany] and UK)
5. Improved worldwide sea-level measurements by more and better controlled tide gauges
6. Improved southern hemisphere GCM's [Global Climate Models] and modeling of ice sheet and ice shelf dynamics
7. a. Paleological, geological, and geochemical determination of possible disappearance of the WAIS during the Eemian interglacial (bottom cores, coral island drilling records, better data, geophysical correlation)
   b. Detailed examination of coral reef cross-sections to determine rate of construction of the 5m Eemian terrace
8. Modeling of inland ice sheet and ice shelf dynamics
   Deep drilling at the south Pole to determine paleo-elevation and paleo-temperatures.[59]

The SIO workshop produced a tighter and more precise report than the wide-ranging scientific presentations at Orono did. The SIO report also mentioned the Eemian period: this would become a critical research question over time, as scientists began to use climate events of the Eemian, the earth's most recent interglacial period, which saw temperatures comparable to what is now predicted for the mid- to late twenty-first century, as an analog for future climate. Perhaps this reflected convergence in priorities as the research community became a little more focused. Or perhaps Revelle, needing information for his upcoming chapter in a major assessment, worked to keep the workshop focused. As host of the workshop as well as the ultimate author of the sea-level rise chapter in *Changing Climate*, Revelle had hand-selected the SIO participants. Debates certainly made it into the formal workshop record, but having fewer participants limited the range of debate. Furthermore, while the Orono workshop seemed to invite participants with conflicting perspectives to spur such debates, the small group of scientists at SIO kept that workshop short, to the point, and with less unwieldy, tangential, or irresolvable discussion.

The WAIS stability problem was mooted again at a climate conference in September 1982 in Berkeley Springs, West Virginia.[60] Here, Bentley's position seemed to harden: he now argued that if there was any evidence of change, WAIS was growing, not shrinking. The ice sheet was "healthy" and he expected it to be around for centuries; even in a warming world it would take 500 years to melt.[61] Drawing on models from Weertman, Budd, and B. J. McInness, Bentley argued that Hughes's estimate that WAIS could collapse in 200 years was wrong. He did allow that a relatively near-term disintegration of WAIS "might barely be possible," but by that he meant a minimum of 500 years.[62]

In response Hughes focused on how to improve and refine WAIS models. He agreed with Bentley that the Weertman and Budd and McInness models had problems, but he nonetheless rebutted Bentley's conclusions. Bentley said that surface melting could not cause rapid disintegration; Hughes disagreed because the meltwater does not stay on the surface of the ice but drains into crevasses near the grounding line and/or ice streams, making it less stable. Bentley also stated that the removal of ice shelves would not cause surges; Hughes again disagreed (and was later proven correct with the 2002 collapse of the Larsen B ice shelf and the resulting surges by several abutting glaciers).[63] Hughes also critiqued a formula that Bentley relied upon; when Hughes ran his correction, he found that the model predicted a rapid disintegration in 200 years. Indeed, Hughes thought WAIS disintegration had already begun and that CO_2-induced warming would accelerate this process.[64] In conclusion, Arnold L. Gordon from Lamont-Doherty submitted his "Comments about the Ocean Role in the Antarctic Glacial Balance," in which he pulled back from the technical arguments between Bentley and Hughes and analyzed the global implications of this debate.[65] The conference report was framed as an adversarial exchange between Bentley and Hughes with Gordon's cooler head encouraging readers to expand the scope of the debate past scientific wrangling and to consider the massive implications if the collapse predicted by some came to pass.

How did Revelle summarize this debate in his chapter in *Changing Climate*? He began by taking a long (100,000-year) perspective on sea level, remarking that "the present is a time of quiet sea level compared with the violent oscillations that occurred during most of the last 100,000 years."[66] Within a shorter time scale, he noted that "the present rate of 10–20 cm per century [of sea level rise] is small compared with the average rate of 1 m per century over the past 15 millennia and very much smaller than the inferred maximum rise of perhaps 5 m per century immediately following the glacial period."[67] The introduction set the tone for a chapter that would imply that while changes were occurring, they were still modest when placed in a longer-term, pre-human civilization context.

Nonetheless, Revelle offered numbers that suggested near-term sea level rise would be neither negligible nor inconsequential: his prediction for global sea level rise through 2100 was about 40 cm through land ice to sea transfer and another 30 cm due to thermal expansion, bringing the total to 70 cm.[68] He gave these estimates large error bars of ±25%, but noted that there were uncertainties that could bring about much more. This included the potential disintegration of WAIS.[69]

To set the local, present-time Antarctic scene, Revelle depended on Bentley and captured the present state of affairs very briefly:

> Estimates of the mass balance of the Antarctic Ice Sheet (Bentley, 1983) suggest that the mass is stable and perhaps even increasing, but the noise level of the estimates is so high that a small net loss corresponding to a rise in sea level of 0.5 mm per year is not forbidden.[70]

Revelle gave WAIS its own section, titled "Possible Disintegration of the West Antarctic Ice Sheet," which served as the capstone for the chapter. Revelle was not shy about the potential five to six meters of sea level rise that might occur:

> The oceans would flood all existing port facilities and other low-lying coastal structures, extensive sections of the heavily farmed and densely populated river deltas of the world, major portions of the state of Florida and Louisiana, and large areas of many of the world's major cities.[71]

The implications for humanity were clear.

Revelle invoked Bentley to characterize the rate and likelihood of WAIS disintegration. As already noted, Bentley, contra Fletcher, had argued that the warming ocean caused thinning at the edges of the ice shelves that buttressed the ice sheet, allowing the upland ice streams to start flowing relatively rapidly into the ocean.[72] Revelle cited Bentley that a discharge spanning 200 years would require "unreasonably high" glacier speeds, though one quarter of WAIS ice could be discharged from Pine Island Bay in that time frame, and half could be in 400 years.[73] According to Bentley's studies, "During this period [of 400 years] the Ross and Filchner-Ronne ice shelves could have disappeared, and all the ice could be discharged within 500 years."[74]

The chapter concluded with a discussion of WAIS research needs. Despite following Bentley in the bulk of the text, here Revelle emphasized that the "disintegration of the West Antarctic Ice Sheet would have such far-reaching consequences that both the possibility of its occurrence and the rate at which disintegration might proceed must thoroughly be researched."[75] The odds might be low, but the consequences were high. Revelle gestured to the Orono workshop recommendations but then spelled out his specific priorities for special emphasis:

> Possible change in the mass balance of the Antarctic Ice Sheet; interaction between the Ross and Filchner-Ronne ice shelves and adjacent ocean waters; ice stream velocities and mass transport into the Amundsen Sea from Pine Island and Thwaites Glaciers; modeling of the ice sheet response to CO_2-induced climate change; and deep coring of the ice sheet to learn whether it in fact disappeared 125,000 years ago.[76]

He also strongly supported satellite-monitoring projects for WAIS and deep-core drilling for both the Greenland Ice Sheet and WAIS.

Revelle's chapter was remarkable in that its tone changed abruptly in the WAIS section. The chapter's opening paragraphs, which put contemporary sea level rise into a global context, were written in a conventional scientific manner. But once Revelle started to write about WAIS, the message was alarming: a veritable collapse of society as we know it. WAIS became the key—a massively uncertain one, but one with tremendous implications.

The *Changing Climate* report was controversial; some of this controversy stemmed from conflict between Revelle and Nierenberg. Revelle's approach, tone, and findings contrasted with Nierenberg's and with the report's executive summary (which Nierenberg had farmed out to NRC staff member Jesse Ausubel). Indeed, the summary ignored Revelle's findings almost entirely. Revelle made a strong argument about a serious threat from WAIS in a warmed world, but the executive summary of the report did not.

US Environmental Protection Agency: 1983

In 1983, the Environmental Protection Agency (EPA) published *Projecting Future Sea Level Rise: Methodology, Estimates to the Year 2100, and Research Needs*, an assessment chaired by John Hoffman (EPA), Dale Keyes (a consultant), and James Titus (EPA). The assessment included contributions from Sergej Lebedeff, Gary Russell, Andrew Lacis, and James Hansen of the Goddard Institute for Space Studies; Robert Thomas and David Thompson of the Jet Propulsion Laboratory; and William Emmanuel of Oak Ridge National

Laboratory. It used a broad, sweeping perspective to introduce the relationship of anthropogenic warming and sea level rise, to discuss its potential impacts on society (with an emphasis on the United States), and to call for further research. It also offered a detailed and nuanced appraisal of the state of the science.[77]

The assessors identified two mechanisms by which global warming's effects on the cryosphere could lead to sea level rise—melting land-based snow and ice; increasing the rate of flow of land-based ice sheets toward the sea—and one by which it might lower it—causing the atmosphere to carry more moisture to cold areas in the form of snow, increasing snowfall accumulation and decreasing sea level. It then provided specific numerical estimates for the potential maximum contributions to sea level rise: 70 meters from Antarctica, 7 meters from land-based ice in the Arctic region including Greenland, and 0.3 meters from high-altitude regions such as the Himalayas.[78] (Their calculations did not separate West and East Antarctica, although a footnote explained that WAIS provided about 10% of the total figure, while the East Antarctic Ice Sheet—thought to be less prone to instability upon warming—made up the remainder.)[79] As in previous assessments, the scientists stressed that a key uncertainty was how ice sheets disintegrate:

> Accurate estimation of the partial deglaciation of the West Antarctic, East Antarctic, and Greenland ice sheets for the next 120 years will require detailed studies of the specific ice sheets. Such studies should consider such factors as the predicted temperature of the upper surface of the ice, surface precipitation rates, ocean water temperatures, melting of ice shelves from the bottom, speeds of ocean currents and their ability to remove ice, the specific topography of the "gates" (narrow areas that constrict the flow of ice), and the specific location of grounding lines (land on which marine ice sheets rest). These factors will determine the speed of discharges. Unfortunately, such studies have not yet been made for deglaciation in the next century.[80]

The concern about the "gates"—such as those found in the Pine Island Bay region—echoed the findings of earlier scientific workshops. The assessors noted that governmental support was completely lacking for this avenue of research: "A comprehensive well-funded effort . . . is still not on the research agenda of any federal agency."[81] Understanding snow and ice transfer (to the sea) was one of the key research priorities; understanding of both the basic physics of ice and models of behavior of the ice sheets as a whole needed to be improved.[82] Despite these uncertainties, the assessors offered numerical estimates for sea level rise by 2100: between 144.4 and 216.6 cm.[83]

Melting ice, including WAIS, contributed heavily to these predictions, but since the rate of melting was so difficult to calculate (because ice melts in a nonlinear fashion), they chose to estimate it as either equal to thermal expansion or twice thermal expansion. These assumptions were supported with model inputs from previously published reports, historical records (noting the problems with continuing historical rates into the future), "process models" (those models containing accurate physics of how ice behaves), and "judgment" (the subjective assessment of the information at hand).[84]

This 144.4–216.6 cm range was extremely high compared to other sea level rise predictions of the time, particularly those offered by Hughes (200 years) and Bentley (500 years) as WAIS deglaciation projections. The authors offered the caveat that "although both estimates were made in the absence of detailed information about melting rates, sea ice retreat, and ocean and air temperatures, the possibility of a complete disintegration in 200 to 500 years cannot be ruled out."[85] The Reagan administration would use this difference to dismiss the EPA report as alarmist and consciously sought to direct attention to the much lower predictions in the *Changing Climate* report.[86] There was now overt political pressure on the scientific community not to be alarmist, but no comparable external pressure not to be complacent.

National Research Council: 1985

In 1985, the NRC published the report of a workshop titled "Glaciers, Ice Sheets, and Sea Level" held in 1984 in Seattle. This was a relatively large workshop, including many key players in WAIS and climate research. The report read like a single-authored book, unified in tone and language. An executive summary was followed by a multiple-chapter report explaining the basic scientific basis, data on sea level changes in the past 100 years, "probable land-ice and ocean exchanges during the next 100 years exclusive of Antarctica," a section on Antarctic work, and a summary with recommendations.[87] Following the formal report, each of the research presentations from the workshop was attached, written up as research articles.

These scientists suggested that there was little if any contribution to sea level rise from the ice sheets at present, but they allowed that in the future that might change. In the executive summary, they wrote:

> The consensus of this Workshop is that sea level is rising, but the rate of rise is uncertain by a factor of 2; wastage of mountain glaciers and small ice caps

> contributes to this rise; probably very little if any sea-level change is caused by wastage of the Greenland Ice Sheet; and the Antarctic Ice Sheet is most likely growing, taking water out of the sea. The rate of change of mass of the ocean cannot be distinguished from zero. Whether the present rise in sea level can be adequately accounted for by just thermal expansion of ocean water is an open question. Future projections suggest that, in spite of increased participation, wastage of small glaciers and the Greenland Ice Sheet will add mass to the ocean; the resulting sea-level rise due to this cause likely will be a few tenths of a meter by year 2100. The sea-level rise due to changes in Antarctica is more uncertain; most likely it will be small, but a rise of an appreciable fraction of a meter by 2100 due to increased discharge of land ice to the sea is not beyond the realm of possibility.[88]

"A few tenths of a meter" by 2100 was much lower than EPA assessment figures. So how had the authors reached their conclusions? Had the science changed that much in two years?

In the report section on Antarctica and sea level change during the past 100 years, the assessors restated their understanding that the continents' ice sheets were virtually within mass balance, neither gaining nor losing much ice. However, in a subsection on the Ross Sea sector of West Antarctica—one of the "best known" areas of WAIS due to extensive measurements taken in the 1970s—they noted that the mass balance of that area "may be slightly positive"—that is, accumulating mass.[89] But, they also noted, "Even in this well-studied region the evidence is equivocal," and "the ice shelf [the floating part of the ice sheet] is capable of noticeable changes in its dynamics on a time scale of a century or two."[90] Moreover, in other areas, they noted, the Antarctic Peninsula was losing mass (but was a small total area of WAIS overall). The Filchner-Ronne Ice Shelf was accumulating at a quite high rate (0.40 meters per year), but the shelf was melting at its base even faster: 1.0 meters per year.[91] And the Pine Island and Thwaites glaciers "may have the potential for rapid ice-sheet disintegration,"[92] since there was little buttressing by the ice shelves at these sites. In other words, the situation was complex and poorly understood:

> There are several research questions that badly need attention if better estimates of present-day Antarctic mass balance and predictions for the future are to be obtained. What are the factors that determine ice-stream flow with sliding? . . . What is the role of the West Antarctic ice streams [fast-flow sections of the ice sheet] in the ice-sheet dynamics; how and why do they play that role? What is the basal mass balance under the ice shelves, and how can it be predicted from oceanographic data? How can changes in thickness of the

> ice shelves be measured separately from measurements of basal melting rates? What is the iceberg calving rate around the Antarctic at present, and how can the effect of a changing environment on the calving rate be determined? What are the time scales for variability in accumulation rate and iceberg discharge?[93]

These questions fell into two general categories: observations needed to improve understanding of contemporary behavior of the ice sheet and physical understanding needed to improve the rudimentary ice sheet models in use at the time. The other subsections of the report also concluded with research needs, albeit not as lists of questions but rather as declarative statements of what measurements were needed or what work should be undertaken in upcoming research trips or satellite missions.

In the chapter devoted to the future of Antarctica, the assessors expanded on the research concerns they had listed as questions. Fundamentally, they argued, there was no way to build reliable ice sheet models if there were no reliable direct observations or models that could mimic direct observations.[94] The chapter authors used Bill Budd's early continental ice sheet models as their premise, particularly when describing the mechanics of the slowly sliding sections of WAIS ice.[95] They used the ice stream models of D. R. MacAyeal and R. H. Thomas, Fastook, and Lingle to describe the more dynamic ice stream behavior.[96] These models represented competing frameworks for estimating the potential rate of disintegration of the ice sheet. There were at that time too few observations to determine which behavior, fast or slow, would dominate loss of ice to the sea as the world warmed. In their analysis, the authors made apparent that these models helped the glaciologists work through some of the key questions about WAIS but that these questions were far from settled. The authors wrote:

> It is clear, then, that we make no *prediction* of a rapid rise in sea level. It is possible that accelerated discharge of land ice from Antarctic will cause a modest rise in sea level in the next century. It is also possible that any accelerated discharge will be offset by increased snowfall on the continent, leading to a small contribution of sea-level change that could either be positive or negative.[97]

Both of these possibilities were modest and small scale, but the first noted the possible significant global impacts that could occur due to WAIS. This muted tone contrasted with the concerned tone of Revelle's *Changing Climate* chapter.

Because the assessors could not state with confidence what sea level rise might occur, they offered a set of tables predicting sea level rise based on various sources and under various scenarios. Most significant, the assessors tried to make numerical estimates of sea level rise "by ice wastage" (i.e., melting and iceberg formation) under climate change. In these, the estimates for the Antarctic Ice Sheet were particularly unwieldy. Both negative and positive contributions were noted, both under conditions identical to the present moment and in 100 years (with increasing levels of CO_2). Following the table, the workshop participants wrote a section on future research needs. This began with a general list: improve climate models, determine the present-day global change in sea level more precisely, and determine to what extent the rise in sea level is due to volume expansion of the ocean.[98] They underscored their general list with more specific recommendations that were expanded upon in great detail. In descending importance: investigate Southern Ocean circulation near Antarctica, ocean/ice shelf interactions, ice streams, and detection and prediction of future changes.[99] To reduce key uncertainties, there would have to be substantial additional research in West Antarctica.

In these six years (1980–1985) of workshops and conferences, we see an ebb and flow in the debates over the future of WAIS. In some contexts, scientists' views seemed to have been relatively consistent, while in other contexts they seemed to be divergent. However, most experts during this period did not view rapid WAIS disintegration as likely to happen anytime in the next few centuries. The exceptions were Mercer, Revelle, and Hughes, whose different views kept the discussion alive and helped to justify more research.

EUROPEAN ASSESSMENTS

International Institute for Applied Systems Analysis: 1981

In 1981, the International Institute for Applied Systems Analysis (IIASA), based in Austria, published *Life on a Warmer Earth*. This report was based on research conducted by Hermann Flohn on the "interaction between energy and climate," jointly funded by IIASA and the United Nations Environment Programme. Flohn was a leading German climatologist who had helped to develop the theory of atmospheric circulation. The IIASA report offered a different story of the origin of interest in anthropogenic climate change than the one typically told about the Swedish scientist Svante Arrhenius. Instead, it focused on Arrhenius's American contemporary Thomas Chamberlin:[100]

> The interest in global warming, like the problem itself, simmered awhile before boiling. In 1899, the American geologist Thomas Chamberlin sounded an early

> alarm. A frequent contributor to the *Journal of Geology*—best known for his hypothesis that the planets had spun off from the sun—Chamberlin attempted in one article to identify an atmospheric basis for glacial epochs and noted almost in passing that carbon dioxide released in the process of burning fossil fuels could warm the lower atmosphere and the surface of the earth.[101]

The authors' history was flawed—Chamberlin's planetesimal theory was the opposite of claiming that the planets had spun off from the sun—but they were right that the possibility of anthropogenic climate change had been known for some time.

Using evidence from the paleoclimatic record, the report contained plentiful discussion of past ice accumulation and deglaciation in Antarctica, although the paper was not focused on only WAIS or even the Antarctic. The authors noted:

> The formation and melting of polar sea ice is a major internal climatogenic process. Historical variations in Arctic sea ice and in the Antarctic continental ice sheet have been considerable. The Antarctic ice is now probably partially unstable, which eventually could cause catastrophic ice surges or deglaciation of western Antarctica. Global atmospheric circulation is asymmetrical because of the contrast between the Arctic ocean's thin cover of drift ice and the Antarctic continent's heavy glaciation.[102]

The report suggested that there was time for five to ten more years of vigorous research before policy action had to be taken and also suggested a CO_2 threshold for disintegration of WAIS at 550–750 ppm. Discussing how long the polar ice caps had been present and at what rough intervals the ice sheets had expanded and contracted, they noted that the West Antarctic Ice Sheet appeared to be different from and less stable than the East Antarctic Ice Sheet:

> A central core of Arctic Ocean [sea] ice has existed for at least the last 700,000 years and probably for more than 2 million years. Since ice caps became well established at both poles, they have expanded and contracted in a strikingly regular pattern. Over the last million years, northern hemisphere ice has peaked roughly every 100,000 years, and the western Antarctic ice sheet has generally kept pace. By contrast, ice sheet changes in the eastern Antarctic have been relatively minor, and they may not have coincided with northern hemisphere glaciations.[103]

Given this, the East Antarctic Ice Sheet was not viewed as a threat, despite its large volume. Instead, it was viewed as the stable, land-based counter to the unpredictable WAIS.

The IIASA assessors were interested in what might cause polar land ice melting and resulting sea level rise, including theories outside the realm of anthropogenic warming. Here is one:

> An unorthodox ice-age hypothesis was proposed in 1964 by A. T. Wilson, a geochemist from New Zealand. He assumed that a combination of pressure from above and geothermal heat flow from below would cause a sufficiently thick body of ice to melt at the bottom. This would make it possible for the Antarctic ice to move forward catastrophically on all sides, forming a quasipermanent ice shelf of 20–30 million km². The result of such a gigantic ice slide would be a general cooling of the earth and the sudden spread of glaciation on the continents of the northern hemisphere. While the bulk of the eastern Antarctic ice is stable and well above sea level (except for a few small meltwater lakes), the smaller ice dome of western Antarctica rests on bedrock below sea level. Indeed this ice may not be stable. It has been suggested that it has disappeared in the geological past, and there is some risk that this could happen again in the foreseeable future.[104]

Wilson's account looked at the ice sheet not only as a feature that could cause global sea level rise but as one that could cool the planet. While it is well documented that atmospheric circulation keeps the cool Antarctic air locked over the continent, Wilson suggested that a displacement of the ice sheet could cause chilled air to spread over a larger region.

The IIASA report summarized the relationship among the ice sheets, the warming earth, and potential sea level rise directly from Flohn's research:

> The amount by which sea levels might rise as a result of global warming can only be put in round numbers. When drift ice melts, sea level stays the same, just as the water line in a drink stays the same as an ice cube melts. Within the next 100 years a significant worldwide rise in sea level could only be caused by large-scale surges of the Antarctic ice cap. As a rough gauge, each 100,000 km³ of ice slide could be expected to raise sea level 25 cm, and a slide of less than 100,000 km³ would probably not affect sea level significantly. The most recent ice slide of sufficient size to raise sea level may have occurred during the last interglacial period, when an Antarctic ice surge of some 2 million km³ could account for a 5-m rise in sea level at that time. Flohn says the risk of such an event recurring as a result of a 4°C increase in global warming is not great during the next century or so. Nor does he consider it likely that the continental ice caps of the Antarctic and of Greenland would melt soon after the North Pole. Even with an ice-free Arctic, Greenland would get more winter snowfall due to increased cyclone activity and probably retain the bulk of its glaciation. But the risks of rising sea level should not be completely disregarded, even though, as

> we have just seen, the risks of a large-scale shift of climatic belts due to global warming are far greater.[105]

On the basis of Flohn's research, the assessors considered sea level rise due to rapid ice sheet disintegration to be a relatively low risk, at least in contrast to the other predicted changes that would be brought about by the forecasted anthropogenic warming over the next century. What is noteworthy, however, is the assessors' willingness to talk about the fate of the ice sheets in the next hundred years. Most reports did not touch 100-year predictions, perhaps because of Bentley's argument that he considered WAIS stable in a 100-year time frame and only conceded a slight possibility of any disintegration whatsoever in a 200-year time frame. Flohn's work suggested that the risk of even a 25-cm increase over the next 100 years due to deglaciation was "not great." The IIASA report demonstrated an interest in idiosyncratic as well as mainstream ideas, showing that assessors were uncertain even about what glaciological theories could be considered fanciful and which were more plausible.

International Energy Agency: 1982

In 1982, International Energy Agency (IEA) Coal Research released *Carbon Dioxide—Emissions and Effects*.[106] The IEA was founded by Organisation for Economic Co-operation and Development member countries in 1974, and it aims to provide critical reviews of interest to "countries interested in minimizing their dependence on imported oil."[107]

The IEA report concluded that CO_2-induced climate change was a major environmental issue that needed more study before effective legal mechanisms could be designed. The authors noted that "because of uncertainties of present knowledge, the development of a management plan for control of CO_2 levels in the atmosphere or of the consequent impacts on society is premature."[108]

The IEA report included a fairly substantial section discussing WAIS in relation to ice and sea level.[109] Despite the group's openly pro-coal bias, their assessment presented a fairly standard overview of WAIS knowledge:

> The question of the disintegration of the West Antarctic ice sheet is still open but it is not likely to surge or breakup in the next century or two. A complete retreat of this ice sheet, which would probably take more than four centuries, would cause a rise in sea level of about 5 m.[110]

By concluding that a disintegration would take over 400 years, the IEA aligned itself closely to Bentley's conservative estimate of 500 years. Overall this assessment took a wait-and-see tone, advocating more, better, clearer, and stronger scientific research before the implementation of climate management policies and actions.

WAIS IN INTERGOVERNMENTAL PANEL ON CLIMATE CHANGE ASSESSMENTS

With the creation of the IPCC in the late 1980s, climate assessments began to shift from being independent, ad hoc affairs to becoming systematic and institutionalized. The first IPCC report has similar language, sources, and general methodological approach to climate assessments as earlier assessments. However, over time, as the IPCC has sharpened its focus and its guidance to authors, the resulting reports have become much more structured and display a higher level of standardization. In many areas, IPCC scientists have also reported increasing confidence in their findings. In particular, IPCC scientists have expressed steadily increasing confidence in their judgment that anthropogenic climate change has occurred. However, this is not the case for their judgments about WAIS.[111] Scientists still struggle to increase their confidence in their understanding of WAIS and its relation to potential sea level rise in a warmer world.[112]

IPCC: The First Assessment Report

In the IPCC's *First Assessment Report* (*FAR*),[113] the assessors used many of the same data and models—or at least updated versions of those data and models—that assessors in the 1980s had used. WAIS was addressed in the chapter on sea level rise (chapter 9), chaired by Richard Warrick, a New Zealander climate modeler, and Johannes Oerlemans, a Dutch glaciologist. Today coordinating lead authors must be nominated by a government, but in the first assessment it was a "different process," as Oerlemans recalled:

> A few scientists were asked [by the IPCC secretariat] if we were willing to prepare a chapter, but the scientists could organize this by themselves. So we could just invite a few people and have a workshop. Now this is all totally different.[114]

To Oerlemans, coordinating the sea level rise chapter for the IPCC assessment was much like coordinating any of the previous sea level rise workshops. In his view that was a good thing:

> I think altogether it's a fairly unique exercise that you do this on a global scale as a United Nations enterprise and—but I must confess that I liked it in the

> beginning more because it was more driven and done by the scientists. It has become much more political, but that is inevitable.[115]

Oerlemans's evaluation of working on the *FAR* compared to later IPCC assessments was that "it was more direct. The process was easier."[116] The initial stage of writing was completed relatively quickly; an argument only broke out once he circulated the draft outside of his workshop. In particular, while 2 of 13 scientists listed as contributors to the chapter were from the United States, Oerlemans recalled that scientists from the United States were not as enthusiastic about the IPCC assessment as those from Europe were. This may have been because the US scientists were conducting their own national climate assessments around the same time, and individuals did not have the time to devote to more than one major assessment. US researchers would later come to play a major role in IPCC assessments, however.[117]

According to Oerlemans, he had invited a prominent American glaciologist to the workshop, but the glaciologist had declined. So Oerlemans sent the American scientist the chapter draft for review, and he received scathing comments in reply:

> He didn't want to come. I wonder if an American was there anyway in this [recording unclear]. I don't think so. They were really missing. Then when we send him out the draft, he came with a huge response and document. It was all wrong and it should be much more about the West Antarctic Ice Sheet. Of course we kind of said, "Yeah. Hello, boy. We will of course look at this but you should realize that you were not there and it would have been so much more easy if you would have attended the workshop," but in the end I'm sure we wrote something about the West Antarctic Ice Sheet.[118]

The authors then passed their draft into the formal review process, which included both governmental reviews and further expert review. In the published product, the authors discussed uncertainties surrounding Antarctic ice sheets this way:

> As to the possible causes and their specific contributions to past sea level rise, the uncertainties are very large, particularly for Antarctica. However, in general it appears that the observed rise can be explained by thermal expansion of the oceans, and by the increased melting of mountain glaciers and the margin of the Greenland ice sheet. From present data it is impossible to judge whether the Antarctic ice sheet as a whole is currently out of balance and is contributing, either positively or negatively, to changes in sea level.[119]

The authors also mentioned rapid disintegration of WAIS as they detailed the "key findings" for their chapter, concluding that a disintegration in the IPCC's primary predictive time scale (to the year 2100) was unlikely:

> On a decadal time scale, the role of the polar ice sheets is expected to be minor, but they contribute substantially to the total uncertainty. Antarctica is expected to contribute negatively to sea level due to increased snow accumulation associated with warming. A rapid disintegration of the West Antarctic Ice Sheet due to global warming is unlikely within the next century.[120]

WAIS also appeared in the report's Summary for Policy Makers, again primarily as a source of uncertainty:

> Although, over the next 100 years, the effect of the Antarctic and Greenland ice sheets is expected to be small, they make a major contribution to the uncertainty in predictions. . . . the West Antarctic Ice Sheet is of special concern. A large portion of it, containing an amount of ice equivalent to about 5 m of global sea level, is grounded far below sea level. There have been suggestions that a sudden outflow of ice might result from global warming and raise sea level quickly and substantially. Recent studies have shown that individual ice streams are changing rapidly on a decade-to-century time-scale; however this is not necessarily related to climate change. Within the next century, it is not likely that there will be a major outflow of ice from West Antarctica due directly to global warming.[121]

While the assessors acknowledged the rapid changes observed in some ice streams, in the absence of convincing data they were reluctant to attribute such changes to a systematic, ice-sheet-scale response to global anthropogenic warming.

IPCC: The Second Assessment Report

In 1996, the IPCC published its *Second Assessment Report* (*SAR*). In the Working Group I Summary for Policy Makers (SPM),[122] the authors stated that "the balance of evidence suggests a discernible human influence on global climate."[123] Using a "business-as-usual" emissions scenario developed by IPCC, the assessors projected a global sea level rise of 50 cm by the year 2100, with a range of 20–86 cm.[124] This estimate was 25% less than the *FAR*'s projection of 66 cm for the same scenario. The *SAR*'s projected temperature increase by 2100 was also lower, which affected the sea level rise projection. After the *FAR*, models had begun to incorporate the effect of particulates in air pollution on earth's temperature, which had the effect of reducing projected warming.

Antarctica received more specific treatment in the *SAR* sea level rise chapter than it had in the *FAR*, while maintaining the consensus that by 2100 the continent's ice sheets would accumulate more ice on a warming earth, because more water would evaporate from the ocean and fall as snow on the ice sheets. The authors also noted that "the observational evidence is insufficient to say with any certainty whether the ice sheet is currently in balance or has increased or decreased in volume over the last 100 years."[125] The polar ice sheets were considered a "major point of uncertainty," and the assessors wrote about the possibility of a WAIS "surge," noting that

> our current lack of knowledge regarding the specific circumstances under which this might occur, either in total or in part, limits the ability to quantify the risk. Nonetheless, the likelihood of a major sea level rise by the year 2100 due to a collapse of the West Antarctic Ice Sheet is considered low.[126]

Very little observational data suggested anything but an ice sheet in balance or accumulating slightly. Later in the chapter, they wrote that "our ignorance of the specific circumstances under which West Antarctica might collapse limits the ability to quantify the risk of such an event occurring, either in total or in part, in the next 100 to 1000 years."[127] However, they believed, "Given our present knowledge" that "if collapse occurs, it will probably be due more to [the residual effect of] climate changes of the last 10,000 years rather than to greenhouse-induced warming."[128] The authors concluded by making a strong plea for satellite altimetry studies of the ice sheet and model improvements.

IPCC: The Third Assessment Report

The 2001 *Third Assessment Report (TAR)* provided a much more detailed and complicated picture of WAIS. In the Summary for Policy Makers of the *TAR* Synthesis Report,[129] the assessors again asserted the likelihood that the Antarctic Ice Sheet as a whole would have a positive mass balance through 2100. However, based on new observational data, they now suggested the Greenland Ice Sheet would lose mass.[130] Projecting beyond 2100, they suggested that "the Antarctic ice sheet is likely to increase in mass during the twenty-first century, but after sustained warming the ice sheet could lose significant mass and contribute several meters to the projected sea-level rise over the next 1,000 years."[131]

The Working Group I Summary for Policy Makers devoted considerable space to outlining future sea level rise scenarios. The paragraph-long caption to its figure 5 began, "The global climate of the 21st century will depend on natural changes and the response of the climate system to human activities,"[132] and stressed the continued uncertainty surrounding potential dynamical changes (i.e., changes in the flow of ice toward the sea) in the WAIS: "Note that the warming and sea level rise from these emissions would continue well beyond 2100. *Also note that this range does not allow for uncertainty relating to ice dynamical changes in the West Antarctic ice sheet,* nor does it account for uncertainties in projecting non-sulphate aerosols and greenhouse gas concentrations."[133] The *TAR* assessors then left rapid twenty-first-century WAIS disintegration and other uncertainties out of their numerical estimates of future sea level rise, a move that would also be made in the fourth assessment.

The *TAR* authors also discussed the prospect for a growing Antarctic Ice Sheet: "The Antarctic ice sheet is likely to gain mass because of greater precipitation, while the Greenland ice sheet is likely to lose mass because the increase in runoff will exceed the precipitation increase."[134] However, even as they told this "positive mass balance" story—one that was scientifically plausible and had been told for decades, partly as a counter to more dire scenarios—it was beginning to be questioned as scientists collected and analyzed more detailed satellite data. The *TAR* authors echoed the broader climate science and policy community's increasing concern over the exclusion of rapid dynamical flow but nonetheless continued to say that rapid loss in the near future was unlikely:

> Concerns have been expressed about the stability of the West Antarctic ice sheet because it is grounded below sea level. However, loss of grounded ice leading to substantial sea level rise from this source is now widely agreed to be very unlikely during the 21st century, although its dynamics are still inadequately understood, especially for projections on longer time-scales.[135]

In this passage, the assessors vascillated between two time scales—the year 2100 mark for which the IPCC assessments try to make predictions and the longer time scale over which anthropogenic climate commitments will occur. While not officially charged to do so, the WAIS assessors in this case decided to make a numerical prediction for the longer-term fate of the ice sheet:

> Current ice dynamic models suggest that the West Antarctic ice sheet could contribute up to 3 metres to sea level rise over the next 1000 years, but such results are strongly dependent on model assumptions regarding climate change scenarios, ice dynamics and other factors.[136]

A three-meter rise would only occur with dynamical ice flow, so the assessors were evidently considering that plausible over the longer time frame. However, even this figure does not point to a total disintegration of WAIS but only a partial—albeit significant—one.

IPCC: The Fourth Assessment Report

In the 2007 *Fourth Assessment Report* (*AR4*), assessors working on WAIS took a step back from the estimates published in the *TAR*, and the numbers they produced were highly controversial. Environmental groups suggested that the lower estimates could be perceived as erroneously minimizing or reversing concern about climate change, and some scientists thought that the implications of the most recent research had not been adequately taken into account.[137]

New research was pouring in that was at odds with the conventional wisdom both of a stable-for-now WAIS and of an Antarctic Ice Sheet with a near-term modestly increasing mass balance.[138] Satellite observations, some based on new instrumentation like the Gravity Recovery and Climate Experiment (GRACE), seemed to suggest that not only was the entire Antarctic Ice Sheet losing mass, but in some places—particularly in West Antarctica and on the Antarctic Peninsula—the loss seemed to be accelerating.[139] Several relatively small floating ice shelves around the Antarctic Peninsula were disintegrating dramatically and rapidly, and the glaciers on the land behind one were accelerating seaward with the loss of the buttressing ice shelf. In addition, new calculations from semi-empirical models that projected sea level rise based on past observations—published after the deadline for inclusion in the assessment—suggested that sea level rise could occur at a rate much faster than other models were projecting, although the ability of these studies to properly account for future ice sheet behavior was questioned.[140]

Besides these new scientific developments, the IPCC Working Group I leadership had decided to try a new approach to characterizing sea level rise. Instead of giving sea level rise its own chapter, aspects of it (and therefore of WAIS) were analyzed in each of four chapters: chapter 4 ("Observations: Changes in Snow, Ice, and Frozen Ground"); chapter 5 ("Observations: Oceanic Climate and Sea Level"), chapter 6 ("Paleoclimate"), and chapter 10 ("Global Climate Projections"). As a result, the past, present, and future of WAIS (and of sea level rise in general) were each discussed separately. This chapter reorganization made it difficult to tell a coherent story about WAIS through time, thus making it difficult to determine what epistemic shifts with regard to WAIS had actually occurred.[141] Indeed, more than one message about WAIS could be found in the report.

In the Summary for Policy Makers discussion entitled "Projections of Future Changes in Climate," the assessors emphasized uncertainty, stressing the lack of knowledge concerning large swaths of basic ice sheet science, which in turn led to serious insufficiencies in their models:

> Models used to date do not include uncertainties in climate-carbon cycle feedback nor do they include the full effects of changes in ice sheet flow, because a basis in published literature is lacking. The projections include a contribution due to increased ice flow from Greenland and Antarctica at the rates observed for 1993 to 2003, but these flow rates could increase or decrease in the future. For example, if this contribution were to grow linearly with global average temperature change, the upper ranges of sea level rise for [the] scenarios shown in Table SPM.3 would increase by 0.1 to 0.2 m. Larger values cannot be excluded, but understanding of these effects is too limited to assess their likelihood or provide a best estimate or an upper bound for sea level rise.[142]

The assessors were hedging: their statement was consistent with the conventional wisdom about WAIS but left open the possibility for the ice sheet to behave in an unpredicted and possibly unpredictable and catastrophic way.

This was an interesting move. The report essentially reiterated the conservative view of WAIS offered in previous IPCC reports, while alluding to the new research that challenged that view. Indeed, a hotly contested debate over WAIS was simmering behind the scenes as the scientists were preparing *AR4*. The wording nodded to both sides of the debate without coming down on either side, although the numbers the authors published ended up consistent with the traditional view. These assessors did similar rhetorical work a few pages later, writing that "current global model studies project that the Antarctic Ice Sheet will remain too cold for widespread surface melting and is expected to gain in mass due to increased snowfall. However, net loss of ice mass could occur if dynamical ice discharge dominates the ice sheet mass balance."[143] The caveats of catastrophic ice loss were tacked onto very incremental ice contribution predictions, so that the *numbers* that were offered were consistent with the conventional view, even while the *words* suggested that something else might occur.

By comparison, in chapter 4 ("Observations: Changes in Snow, Ice, and the Frozen Ground"), Working Group I assessors did more to challenge the conventional view. They noted that at least since 1993 the Antarctic Ice Sheet had been detected to have negative rather than positive mass balance, and they concluded that the Antarctic Ice Sheet, along with the Greenland Ice Sheet, had been losing mass and contributing to global sea level rise:

> Taken together, the ice sheets in Greenland and Antarctica have *very likely* been contributing to sea level rise over 1993 to 2003. Thickening in central regions of Greenland has been more than offset by increased melting near the coast. Flow speed has increased for some Greenland and Antarctic outlet glaciers, which drain ice from the interior. The corresponding increased ice sheet mass loss has often followed thinning, reduction or loss of ice shelves or loss of floating glacier tongues. . . . The recent changes in ice flow are *likely* to be sufficient to explain much or all of the estimated Antarctic mass imbalance, with changes in ice flow, snowfall and melt water runoff sufficient to explain the mass imbalance of Greenland.[144]

In short, the Antarctic and Greenland Ice Sheets had contributed to recent sea level rise, and that contribution seemed to be accelerating. But was this just a short-term fluctuation or the beginning of an ongoing, even rapid, loss of ice to the sea?

The assessors did not answer this question, restating the observed recent accelerations of ice loss as well as the uncertainties surrounding these accelerations; the uncertainties made the future of the ice sheets unknowable and unpredictable at this time:

> Results summarised here indicate that the total cryospheric contribution to sea level change ranged from 0.2 to 1.2 mm yr^{-1} between 1961 and 2003, and from 0.8 to 1.6 mm yr^{-1} between 1993 and 2003. The rate increased over the 1993 to 2003 period primarily due to increasing losses from mountain glaciers and ice caps, from increasing surface melt on the Greenland Ice Sheet and from faster flow of parts of the Greenland and Antarctic Ice Sheets. Estimates of changes in the ice sheets are highly uncertain, and no best estimates are given for their mass losses or gains. However, strictly for the purpose of considering the possible contributions to the sea level budget, a total cryospheric contribution of 1.2 ± 0.4 mm yr^{-1} SLE [sea level equivalent] is estimated for 1993 to 2003 assuming a midpoint mean plus or minus uncertainties and Gaussian error summation.[145]

The assessors struggled to find a way to incorporate the new direct observations showing rapid changes in parts of WAIS, which countered the standard assessment of little to no change for WAIS through 2100:

> The ice sheets of Greenland and Antarctica are the main reservoirs capable of affecting sea level. Ice formed from snowfall spreads under gravity toward the coast, where it melts or calves into the ocean to form icebergs. Until recently (including IPCC 2001) it was assumed that the spreading velocity would not change rapidly, so that impacts of climate change could be estimated primarily from expected changes in snowfall and surface melting. Observations of rapid ice flow changes since IPCC (2001) have complicated this picture, with strong indications that floating ice shelves "regulate" the motion of tributary glaciers, which can accelerate manyfold [*sic*] following ice shelf breakup.[146]

The newer observations of quick-flowing ice streams raised questions about both the models considered most trustworthy and the story that WAIS researchers had been telling for two decades. The Mercer-like "threat" now seemed to be credible, but while there was enough information to hint at a reversal to the story, the assessors did not feel they had enough information or time to rewrite the story by the deadline for the assessment report.

One of the key controversies of WAIS in *AR4* centered around a table that was published in the Synthesis Report and in the Working Group I Summary for Policy Makers.[147] This table showed sea level rise estimates with the caveat "model based range excluding future rapid dynamical changes in ice flow." The authors explained this using bullet points dispersed throughout the Synthesis Report's Summary for Policy Makers. These included

- The projections do not include uncertainties in climate-carbon cycle feedbacks nor the full effects of changes in ice sheet flow, therefore the upper values of the ranges are not to be considered upper bounds for sea level rise. They include a contribution from increased Greenland and Antarctic ice flow at the rates observed for 1993–2003, but this could increase or decrease in the future. {3.2.1}[148]
- Current global model studies project that the Antarctic ice sheet will remain too cold for widespread surface melting and gain mass due to increased snowfall. However, net loss of ice mass could occur if dynamical ice discharge dominates the ice sheet mass balance. {3.2.3}[149]
- *Risks of large-scale singularities.* There is *high confidence* that global warming over many centuries would lead to a sea level rise contribution from thermal expansion alone that is projected to be much larger than observed over the twentieth century, with loss of coastal area and associated impacts. There is better understanding than in the *TAR* that the risk of additional contributions to sea level rise from both the Greenland and possibly Antarctic ice sheets may be larger than projected by ice sheet

> models and could occur on century time scales. This is because ice dynamical processes seen in recent observations but not fully included in ice sheet models assessed in the AR4 could increase the rate of ice loss. {5.2}[150]

Working Group I's Summary for Policy Makers was clear about the confounding effect of the new research on providing assessments for WAIS and ice contributions to sea level rise in general:

> New data since the TAR now show that losses from the ice sheets of Greenland and Antarctica have *very likely* contributed to sea level rise over 1993 to 2003 (see Table SPM.1). Flow speed has increased for some Greenland and Antarctic outlet glaciers, which drain ice from the interior of the ice sheets. The corresponding increased ice sheet mass loss has often followed thinning, reduction or loss of ice shelves or loss of floating glacier tongues. Such dynamical ice loss is sufficient to explain most of the Antarctic net mass loss and approximately half of the Greenland net mass loss. The remainder of the ice loss from Greenland has occurred because losses due to melting have exceeded accumulation due to snowfall. {4.6, 4.8, 5.5}[151]

While the assessors felt confident in noting the shifts in recent research, they felt unable to translate these shifts into numerical predictions about rapid dynamical ice flow in the near- or long-term future.

Our data collection and period of analysis ended with the *Fourth Assessment Report*. However, in its 2014 *Fifth Assessment Report* (*AR5*), IPCC revisited these questions but changed its approach. Based on the outcomes of improved modeling and additional observations of ice sheets and sea level, it projected the contribution of Antarctic ice flow to twenty-first-century sea level rise for the first time. That and other adjustments led to an increase of about 60% in estimated sea level rise by 2100. More research is needed to understand how assessors chose to calculate sea level rise estimates in *AR5*, but we notice both a shift toward diverse types of models and more reliance on expert judgment along with continuing hesitation to fully assess the question of rapid disintegration in light of its scientific uncertainties.

CONCLUSION

Why did scientists involved in assessing WAIS believe for so long that it was stable and represented no policy-relevant threat? One answer to this question involves the problem of consensus, or what we call "univocality." Scientists participating in assessments often feel the need to achieve agreement in order to send a clear message to policy makers.

This offers a significant point of contrast with science undertaken in academic contexts: in the latter, there is no particular pressure to achieve agreement at any given moment or by any particular deadline (except perhaps within a lab group, in order to be able to publish findings or write a grant proposal). Moreover, in academic life scientists garner attention for their work and sometimes prestige by disagreeing with their colleagues, particularly if the latter are prominent. The reward structure of academic life leans toward criticism and dissent; the demands of assessment push toward agreement. These differences in reward structure may have epistemic consequences.

The impulse toward univocality arose strongly in debates over how to characterize the risk of rapid disintegration of WAIS. Scientists believed that it was important for them to come to some kind of agreement—even a very modest one—despite substantial differences of opinion about the character and imminence of that risk. This led to a "least common denominator" finding, the minimum conclusion that everyone could agree upon. For WAIS, it was the conclusion that it would lose most of its ice in the very long run with sufficient warming (nearly all experts agreed on this) but not in the short run (because experts did not agree on that). This outcome was "conservative"—where "conservative" was defined by scientists as under- rather than overestimating the threat. In hindsight it appears to have been incorrect.

The pressure for univocality need not necessarily lead to an underestimation of a threat, but in this case it is clear why it did. Everyone agreed WAIS might disintegrate in 500 years or more; not everyone agreed that it could disintegrate sooner. So the area of overlap was on the "less worried" side. One could imagine a situation where the opposite was true, where scientists overlapped on the more worried side. But there is more to the problem than that.

Elsewhere we have documented a pattern we label "erring on the side of least drama."[152] By this we mean that scientists often have a tendency to avoid dramatic findings, because drama is associated with emotion, feelings, irrationality, and even femininity, qualities that have traditionally been viewed as at odds with scientific rationality. We have shown that in several domains related to climate change, scientists' estimates of various threats—CO_2 emissions, Arctic sea ice loss, sea level rise—have tended to be low relative to actual outcomes. Scientists considered such underestimates to be "conservative" because they are conservative with respect to the question of when to sound an alarm or how loudly to sound it. (It is of course not conservative when viewed in terms of giving people adequate time to prepare.) The history recounted here is consistent with this finding: that WAIS assessments underestimated the threat of rapid ice sheet disintegration, because most of the scientists who participated were more comfortable with an estimate that they viewed as "conservative" than with one that was not.

The history of WAIS assessments also illustrates how the demand for assessment can shape a research agenda. Throughout the history recounted here, we see scientists consciously defining research agendas in terms of what needs to be known in order to answer what is perceived to be the key policy-relevant question: Is the WAIS at risk of rapid disintegration in a policy-relevant time frame? Many of the topics addressed would no doubt have been of interest to glaciologists with or without the pressures and incentives of assessment, but one element that was clearly foregrounded, and might not otherwise have risen to such prominence, was ice sheet modeling. While many aspects of WAIS could be better understood through observational studies, the demand to be able to predict its future behavior required modeling.

Modeling became widespread throughout the earth sciences during the period considered here, but a particularly strong preference for computer models of physical (and sometimes social) processes has become institutionalized, especially within the IPCC, as both scientists and governments have devoted increasing resources to developing and running processed-based models to be used in IPCC assessments. These models are unquestionably valuable; the question is, what happens when reliable model-based results are not available? This occurred in the assessment of the risk of sea level rise from rapid WAIS disintegration in *AR4*. Lacking results from process-based models in which scientists had confidence, they left out of their highlighted numerical predictions any potential increased contribution from rapid WAIS integration, in effect assigning it a value of zero.[153] This occurred despite the fact that other estimation approaches were suggested by some participants.[154] The result was that the assessment offered numbers that many scientists acknowledged, even at the time, were at best incomplete and at worst severely misleading. The post-*AR4* spurt of research served as a midcourse correction, and *AR5* succeeded in reducing this bias up to a point. But some scientists think the numbers in *AR5* were still too low. And what other aspects of IPCC assessment have been subject to similarly biased treatment?

The history of WAIS assessments also illustrates the anchoring effect of early assumptions and conclusions. The narrative arc of WAIS since the 1950s starts with concerns over stability: It was understood that WAIS was a marine ice sheet, grounded below sea level, which might make the ice sheet structurally unstable. However, despite this understanding, the earliest WAIS glaciologists contended that their observations indicated that WAIS would be stable at least for the next several centuries. Early three-dimensional ice sheet models, such as the one developed by Philippe Huybrechts and Johannes Oerlemans, considered the best available through the early 2000s, supported these claims.[155]

However, expert perspectives on WAIS stability were never uniform. For as long as Charles Bentley's work suggested a stable ice sheet, other scientists disagreed. Some glaciologists studying more dynamic areas of the ice sheet were more sensitive to potential instability and to the possibility of greater sea level rise. Other scientists may simply have been playing the role of devil's advocate.[156] But, as we have seen, the "conservative" position, wherein "conservative" is understood to be equivalent to reassuring, largely prevailed.

Eventually, though, observations started to shift in favor of the ice sheet's near-term instability, and in the run-up to *AR4* a great deal of new information from satellite observations and other sources showed that the standard ice sheet model was seriously deficient. This upended the consensus for WAIS stability. Yet the numerical values that were put forward for WAIS contribution to sea level rise did not reflect this new situation. The conservative position had had an anchoring effect; it took yet another assessment (*AR5*) to dislodge it.

Finally, we may note a change in the character of WAIS assessments, consistent with the point made in chapter 1 that assessments for policy in the late twentieth century are distinctively characterized (in contrast to earlier reviews and commissions) by their institutionalization. WAIS assessments began as cozy and collegial ad hoc weekends in retreat centers; these gave way to workshops that included multiple drafts and review comments and then to IPCC assessment with rather formalized (although not entirely formal) procedures. Among other things, these procedures ensure that a wider range of scientists is involved, including more government scientists, more scientists from around the globe, and more women. Procedural rules may also help to ensure continuity, diversity, a more consistent depiction of scientific uncertainty, and some degree of transparency. Accompanying their institutionalization, assessments now involve a much greater expenditure of person-hours and money than they did in the past. Whether that increase in expenditure has been worthwhile depends in part on what effect recent assessments have on public policy, and that remains an open question.

WAIS TIMELINE

1957–1958

- Glaciologists begin multiyear research on WAIS as part of International Geophysical Year

1958

- Scientific Committee on Antarctic Research established

1968

- Paper by John Mercer links atmospheric warming with potential rapid disintegration of WAIS

1978

- Mercer's "Threat of Disaster" paper published in *Nature*

1980

- April: Orono Conference

1981

- IIASA report *Life on a Warmer Earth* published

1982

- February: "CO_2, The West Antarctic Ice Sheet, and Global Sea Level" workshop at SIO
- September: Carbon Dioxide Research Conference in Berkeley Springs, West Virginia
- IEA Coal Research report *Carbon Dioxide—Emissions and Effects* released

1983

- US *Changing Climate* assessment published
- EPA *Projecting Future Sea Level Rise* assessment published

1985

- NRC report of 1984 Seattle workshop on "Glaciers, Ice Sheets, and Sea Level" published

1990

- IPCC *First Assessment Report* (*FAR*) published

1995

- IPCC *Second Assessment Report* (*SAR*) published

2001

- IPCC *Third Assessment Report* (*TAR*) published

2007

- IPCC *Fourth Assessment Report* (*AR4*) published

2014

- IPCC *Fifth Assessment Report* (*AR5*) published

CHAPTER FIVE

Patrolling the Science/Policy Border

INTRODUCTION

We have seen that the participants in our case studies view the character of the relationships between science and policy—and the borders between them—as a very important matter. In theory, the relationship between these domains is clear: scientists, acting as independent professional experts, self-assess their knowledge base and make this information available to those who may wish to use it (or not) to inform policy choices. In doing so, they enact the supposed fact/value distinction, which in its modern form goes back to the eighteenth-century philosopher David Hume and was developed in the nineteenth and twentieth centuries by the positivist tradition in philosophy of science.[1] Scientists, with their technical knowledge, are experts about facts, but the value decisions implicit in policy choices involve considerations that extend beyond their expertise. For all their technical competence, natural scientists lack the training and perspective to be authorities in these value-laden domains, and in democratic societies it would be inappropriate for them to usurp the role of policy makers. However, in the practice of assessment, there are no absolute (or even consistent relative) standards for the relationships between facts and values, science and policy, and the technical and the political.

In the assessments discussed in this book, scientists frequently discussed—and in some cases disagreed about—the relations between science and policy. In chapter 2, we saw that the US National Acid Precipitation Assessment Program (NAPAP) was mandated by Congress to "identify actions to limit or ameliorate the harmful effects of acid precipitation."[2] Formally, policy recommendations were to be part of the assessment. However, in the hostile political environment of the Reagan administration, NAPAP scientists were reluctant to engage with the development of policy options and attempted to draw firm boundaries between the assessment of the state of scientific knowledge and policy recommendations—mainly by focusing on the former and postponing the development of the latter. Yet, as some participants noted, the development of scenarios at the end of the 10-year program blurred this boundary or even effaced it. Value-laden decisions were made, for example, about what the future energy system would look like.

In chapter 3 we saw that stratospheric ozone assessments occurred in various institutional contexts and were governed by different charges, but most participants agreed that it was crucial to distinguish scientific findings from policy recommendations. A range of strategies were employed to do this, including scenario development, dividing reports into science and policy sections, and, in one case, commissioning two separate but largely simultaneous reports written by committees with overlapping staff and membership.

Participants in the assessments that we studied often stressed the importance of preventing the seepage of political considerations into their work, holding (sometimes with great vehemence) that scientists should tell governments what is the case but not presume to tell them what do to about it. (One is reminded of Galileo's famous quip that scientists may know how the heavens go but not how to go to heaven.) When asked why this is important, scientists often framed their answers in terms of neutrality and credibility: if scientists make policy recommendations, this will be viewed as compromising their (perceived) neutrality and therefore their objectivity, which in turn will result in loss of credibility. In other words, they believe they must be policy-neutral to be objective and must be objective to be credible and effective as advisers. Therefore, they must refrain from making policy recommendations.

Over time, ozone scientists came to view their role as providing "policy-relevant but not policy-prescriptive" information, a distinction that lives on explicitly in the policies and practices of the Intergovernmental Panel on Climate Change (IPCC).[3] But a pithy slogan does not in itself constitute a solution to a problem.

Policy decisions are rarely binary choices. Typically, they involve a portfolio of possible interventions, each of which can be implemented to a degree, but none of which is fully separable from technical information and the way in which it is framed. For example, if our goal is to protect lakes, then we need to cut sulfur emissions enough to ensure that goal is reached, and this requires scientific knowledge to help determine how to do this and what level of reduction counts as protection.

Consider another example. The objective of the United Nations Framework Convention on Climate Change (UNFCCC) is to avoid "dangerous anthropogenic interference with the climate system."[4] Policy makers need scientists not only to help identify means for reaching this goal but also to help determine the level at which danger ensues. This need—for technical information to support a piece of governance—has over time led to the widely accepted view, now embedded in the Paris Agreement, that the world must keep the increase in global average temperature to well below 2°C in order to meet the UNFCCC objective.[5] But 2°C is no magic number; it is not a physical constant like Avogadro's number or the acceleration of gravity. It has no inherent physical significance. (Indeed, the Paris Agreement suggests that 1.5°C may be a more appropriate objective, and in principle one could propose 1° or even 0°.) If the West Antarctic Ice Sheet has already become unstable (as some scientists think), then "dangerous anthropogenic interference" may already have occurred.[6] In any case what constitutes "dangerous" goes beyond emissions profiles and the physics of ice sheets and involves social, economic, and biological facts that bear on not only disruptive effects but also the capacity to adapt. The choice of 2°C as the marker of "dangerous anthropogenic interference" emerged from a complex process in which science and policy mutually informed (some would say deformed) each other.[7]

It is generally accepted, at least with respect to the issues that we discuss in this book, that the policy process should be informed by scientific information, but less accepted and understood is the extent to which scientific work in assessments is itself informed by policy processes. The mutual influence between science and policy arises in part because of the indeterminacy of the border between them. Some issues that may look to an outsider like policy matters may be considered by scientists to be amenable to technical analysis, and some matters that scientists wish to avoid as "political" might seem to a layperson to be highly technical. Moreover, discussion of policy goals is sometimes explicitly part of the assessment or its charge (e.g., NAPAP and some ozone assessments). Some social scientists would argue further that scientific assessments by their very nature are part of a policy process, and scientists involved in assessments are for that reason involved in policy construction.[8] By participating in an assessment, a scientist affirms at least to some degree that a particular activity or set of activities has created a problem: that it is "a matter of concern."[9] Acid rain, ozone depletion, and sea level rise became topics of assessments only after sufficient evidence had accumulated for scientific experts to say that these matters presented at least a potential threat. In addition, some scientists who take part in assessments are sentinels, calling attention to an issue such as acid precipitation or ozone depletion about which we might otherwise not be concerned or even aware. The sentinel warns of impending danger—but danger is not a scientific concept but a social one, involving notions of harm, injury, peril, and menace.

Scientists' views of their appropriate role in addressing policy questions have changed over the course of the past half century. In the 1950s and 1960s, many prominent scientists felt it urgent to speak up on pressing matters of concern and felt justified in recommending remedies. Nuclear physicists, most famously Niels Bohr, Albert Einstein, and Hans Bethe, spoke out vigorously on the threat of nuclear weapons and the need for arms control to address the threat. Biologists Garrett Hardin, Rachel Carson, Paul Ehrlich, and Barry Commoner addressed the threats of environmental pollution; Hardin and Ehrlich became aggressive public advocates of population control.[10] Why did these scientists consider it appropriate not only to speak up publicly but also to offer solutions? Why did scientists in several recent assessments come to take a different view, concluding for the most part that they should refrain from making policy recommendations? And how have scientists in assessments attempted to build and sustain a recognizable boundary between the "policy relevant" and the "policy prescriptive"?

In order to understand how scientists in assessments have come to their present views, we step back and consider the scientists in the mid-twentieth century who embraced the dual role of sentinel and problem solver.

SCIENCE AND POLICY: AN UNCERTAIN RELATIONSHIP

In the years immediately following World War II, a number of prominent physicists began to speak publicly on the threats represented by nuclear warfare and the risks of a nuclear arms race. The most famous was Niels Bohr. Even before the end of World War II, Bohr began to speak strongly of the urgent need created by nuclear weapons for international cooperation to control their spread.[11] At first Bohr tried to communicate privately to political leaders; later he reached out to the United Nations as a political entity that might address the threat. Albert Einstein embraced a public role, speaking out during the war against the Nazi threat and after the war for arms control (and in later years for Zionism, pacifism, socialism, and civil rights, and against McCarthyism). Both Bohr and Einstein became public intellectuals speaking on diverse topics, many related to their expertise in matters nuclear, but others not. While their interventions were not uniformly welcomed (after his meeting with US President Franklin Roosevelt in 1944, Bohr's loyalties were questioned, and the US government limited his participation in the Manhattan Project and placed him under FBI surveillance), these men were widely sought as public speakers and commentators, their views widely cited in mass media.[12] Einstein, famously, became a celebrity, the iconic genius of the twentieth century.

Media coverage of these men's views suggested that they were of more than ordinary value, in light of the exceptional brilliance of the men who held them. Their voices were taken to reflect the insights of science, perhaps even the deliverances of reason.[13] But their own argument was that they spoke not as geniuses but as experts—specifically physicists—who had a particular understanding of the threat that nuclear weapons present. Their view was that the US government, and the public at large, needed physicists to impress upon them just how very concerning nuclear weapons were. They were soon joined in this sentinel role by other leading nuclear scientists, including Robert Oppenheimer, Hans Bethe, George and Vera Kistiakowsky, and (in France) Frederic and Irene Joliot-Curie.[14]

Some scientists had spoken up against the atomic bomb even before World War II was over. In 1945, Leo Szilard, the Hungarian-born scientist who wrote the "Einstein letter" urging President Roosevelt to begin a project to build an atomic bomb, began a petition drive at Los Alamos against the use of the bomb. Scientists at the University of Chicago led by physicist James Franck urged the US government not to use the atomic bomb without first inviting the Japanese to a test demonstration.[15] Franck argued that scientists' intimate involvement in the question of atomic weaponry, including their "prolonged preoccupation with its world-wide political implications," not only justified but imposed upon them the obligation to offer their views on how to control them.[16]

In the immediate postwar period, when nuclear weapons were new and few, Szilard, Franck, and others argued that physicists were in a unique position to understand the threat, communicate it, and protect the world from annihilation. Soon many US scientists began to argue something close to Szilard and Franck's position: that they had an active responsibility to engage in discussions of the bomb's future not only by virtue of their role in building it but also by virtue of their intimate knowledge of and proximity to the problem. The so-called Scientists' Movement—initially an informal assortment of voices but later organized into the Federation of Atomic Scientists and then renamed the Federation of American Scientists—held that their familiarity with atomic weapons gave them a particular, specific, and immediate responsibility to engage in public discussion of them.[17] The environment of their interventions was historically unique, and for some it was this uniqueness that warranted their outspokenness. For others, it was a sense of special responsibility as scientists.[18]

Yet physicists were by no means unanimous about their role. Robert Oppenheimer, chief scientist on the Manhattan Project, opposed the Szilard petition on the grounds that the bomb's use was outside their domain of expertise.[19] A young Richard Feynman went further, claiming to practice "active irresponsibility" as a matter of principle—indicating that he cared not what the consequences or applications of his scientific work were, nor should he.[20] President Harry Truman agreed that scientists were not the appropriate experts to make the decision whether to use the bomb: he ignored their opposition and, following the advice of his political and military advisors, used it against civilian targets in Hiroshima and Nagasaki with no advance notice.[21]

Oppenheimer's opposition to the Szilard petition may have reflected his sense that it was his job to build the bomb, not to proffer advice on its use. But after the war, he began to change his views, allowing that it was "true that we are among the few citizens who have had occasion to give thoughtful consideration to these problems."[22] In 1946, he would be coauthor, along with other Manhattan Project luminaries Hans Bethe, Arthur Compton, Walter Alvarez, and Glenn Seaborg, of the *Report on the International Control of Atomic Weapons*—known as the Acheson-Lilienthal report for the chairs of the committee that produced it—which advocated international control of fissile materials.[23] The justification that these men offered for this foray into the political realm was the same that Szilard and Franck had offered: their intimate scientific knowledge of nuclear weapons gave them a particular—even unique—appreciation of the political and existential threat that these weapons represented. Scientists also insisted that the framework of protecting and maintaining "atomic secrets" was implausible; Soviet scientists would catch up quickly if they had not already.[24] Even scientists like Seaborg who supported expanded weapons development nevertheless agreed that scientists should play a leading role in the control of fissile materials.

When the time came to consider the hydrogen bomb a few years later, Oppenheimer and his colleagues dove deeply into an argument that was neither primarily scientific nor technical. Leading physicists (including Oppenheimer) initially opposed the H-bomb on moral grounds: asked in 1949 whether the H-bomb should be built, a majority of the General Advisory Committee to President Truman, whose members were scientists, said no, because any use would necessarily kill civilians in copious numbers. A minority of the committee—including Enrico Fermi—went further, arguing that, as a genocide weapon, it was "necessarily an evil thing considered in any light."[25]

This foray into morality had a cost. After the American decision to build the H-bomb was made, Robert Oppenheimer was humiliated and stripped of his security clearance, at least in part because of his opposition.[26] Most leading scientists—including the conservative stalwart Vannevar Bush—defended Oppenheimer, but for scientists unsure of where or even whether they belonged on the ship of state, it was a clear shot across the bow.[27] Many scientists were chastened by this episode and saw that reticence on policy questions was a safer strategy than candor.

The Rise of Select Committees

The US government recognized that it needed scientific advice, even if it did not like all the advice it received. Especially after the Soviet Union placed the first artificial satellite into space (Sputnik in 1957), the US government increasingly turned to scientists for help in prosecuting the Cold War. The need for advice on diverse technical questions was formally recognized by the creation of new institutional structures, most notably the President's Science Advisory Committee (PSAC)—a strengthened and upgraded form of the World War II–era Science Advisory Committee—and the "Jasons," a secretive group of scientists that advised the US Department of Defense and the Atomic Energy Commission (later the Department of Energy) throughout the Cold War (and continues to do so today).[28] Similar institutional structures were created in other countries.[29]

Despite Oppenheimer's downfall, the scientists involved in these committees tended to define their role expansively, although PSAC members were mindful of the need not to overstep their authority, especially since their responsibility was explicitly to advise the president of the United States (the famous question of whether scientists should be "on top" or "on tap").[30] Many of the questions they addressed were not exclusively technical. Moreover, the argument was made—most notably by a president, Dwight Eisenhower—that the distinction between science and policy was too crudely wrought. President Eisenhower felt that his advisors would be more useful to him if they could "liberate themselves from their 'exact' mind-set to see beyond the logic of technological determinism and take political factors into considerations in the policy realm."[31] Technical arguments, the president believed, should be balanced with "those derived from other justified sources." The president wanted his advisors to take these other sources seriously and offer a full range of views.[32]

PSAC in its day made many policy recommendations. The committee supported the Limited Test Ban Treaty and civilian control of the US space program against military opposition. Committee members criticized what they perceived to be misguided military projects, such as aircraft nuclear propulsion, and gave extensive advice about weapons systems and alleged Soviet missile number superiority.[33] In these areas, PSAC scientists agreed, technical and political considerations were closely linked, if not inseparable. Discussing a review of the 1961 Department of Defense budget, for example, they wrote: "We have not found it possible to limit our review to purely technical considerations in view of the complex interaction between weapons technology and non-technical factors."[34] Meanwhile, scientists like Herbert York, the first director of defense research and engineering at the Pentagon, joined with PSAC members in arguing against "technological palliatives to cover over serious persistent underlying political and social problems."[35] York was not alone in becoming an advocate for arms control in light of the futility of trying to solve the problems posed by nuclear weapons by building more of them, and President Eisenhower supported his approach.[36]

The overarching philosophy of PSAC in the Eisenhower administration was that "technical issues could never be neatly and completely separated from social, economic, and political factors, and what was technically feasible was not always desirable."[37] PSAC's impact derived from this recognition and from its willingness to go beyond the narrowly technical in its analyses.[38] In the words of the nation's first science advisor, James Killian, the "scientific" issues they addressed "involve political, ethical, and scientific considerations in a way that . . . cannot be wholly disentangled."[39] This overall philosophy continued into the Kennedy and Johnson administrations. More focused on domestic policy than his predecessors, President Johnson particularly wanted "scientists to help make life better for 'grandma.'"[40]

This is not to say that PSAC scientists never attempted to draw lines between science and policy; at times they did. But they understood their role to be both scientific and political and believed that artificial distinctions between these domains could lead to flawed analyses and costly errors.[41]

In the 1960s and 1970s the Jasons were almost entirely physicists (though they have become somewhat more diverse in recent years), but that did not prevent them from giving advice when asked about military policy in Vietnam, the desirability of building a supersonic transport (SST) fleet, negotiating an antiballistic missile (ABM) treaty, and whether climate change was something to worry about. Much like the scientists who opposed developing the H-bomb in the 1950s, they argued against carpet bombing in Vietnam on moral grounds.[42]

Before long PSAC would be accused of overstepping, precipitated by committee member Richard Garwin's opposition to President Richard Nixon on the question of ballistic missile defense. PSAC had also opposed ABMs under Johnson—on political as well as technical grounds—a position that was adamantly rejected by Johnson's Joint Chiefs of Staff but supported by Defense Secretary Robert McNamara. Johnson did not have strong views on ABMs, so the disagreement did not escalate or become public. Nixon, however, wanted an ABM system, and the public stance against them taken by some PSAC members, as well as their publicly expressed opposition to the Vietnam War, angered Nixon.[43] This anger was compounded when Garwin testified in Congress against the SST program, which Nixon favored. Garwin's public opposition to the president's policy led Nixon in 1973 to dissolve PSAC.

The importance of appointed committees and panels should not obscure other ways in which scientists made their voices heard. While physicists were largely giving advice through formal channels, scientists in other fields continued to play the role of sentinel informally, particularly in the emerging arenas of environmental science and population. In 1962, Rachel Carson's best-selling book *Silent Spring* thrust her into the limelight as Americans became aware of the risks posed by widely used pesticides.[44] Biologists Paul Ehrlich and Garrett Hardin became public figures as they linked environmental damage to population growth.[45] Barry Commoner's best-selling book *The Closing Circle* tied environmental destruction to wasteful industrialism, consumerism, and forms of governance that did not hold those responsible accountable.[46]

Meanwhile, physicists continued to advocate for and participate in arms control negotiations and agreements. In the early 1970s, Wolfgang Panofsky, director emeritus of the Stanford Linear Accelerator Center and member of the US National Academy of Sciences Committee on International Security and Arms Control, helped to open a back channel to the Chinese government through his contacts in the Chinese physics community.[47] In the fraught domain of arms control, personal relationships seem to have played an important role in building relations of trust. Elsewhere, for example, in the domain of environmental policy, elite committees and informal personal approaches were being overshadowed by the rise of organized assessments of science for policy.

The Rise of Organized Assessments

The rise of the formalized assessments that are the subject of this book does not quite coincide with the fall of PSAC, but it comes close: NAPAP was authorized by Congress in 1980; the first successful international ozone assessments were published in 1982 and 1986; the IPCC was created in 1988. If we take the end of the Cold War as a marker, a distinctive feature of the post–Cold War period is the existence of very large, international, and highly bureaucratic forms of scientific advice. Scientific advice for policy is no longer being offered primarily by celebrated individuals, famous for their scientific contributions, or by small groups of distinguished invited scientists (who may also have been asked for their policy recommendations), but by large groups of scientists—scores, hundreds, or even thousands. Few of them are known to the public; most may be fairly described as "rank-and-file" scientists. So whereas in the mid-twentieth century scientific advice for policy was mostly offered by famous individuals working on their own accord and then for a time by select committees of hand-picked men (and the very rare woman) serving national governments, beginning in the mid- to late twentieth century large formal assessments, typically international, have become the dominant source of advice.

One reason given by scientific participants for the need for international assessments was that they came to believe that international assessments would be viewed as more objective—and would therefore carry more authority—than national assessments that would be perceived as tethered to the policy aims of the governments of the countries involved.[48] This can be seen as part of a broad strategy of seeking to construct externally evident markers of objectivity, an important piece of which was to demarcate technical answers as distinct from policy recommendations.[49]

Like any periodization, the one offered here should not be too sharply drawn. During any of these periods, we can observe a spectrum of behavior and rationales. However, we do find that the scientists in the assessments studied here have broadly concluded that external markers of objectivity matter, particularly the demarcation of the technical from the political. These scientists also generally affirm that they should hew to the technical side and not offer policy advice. Given the evidence that leading scientists in the 1950s, '60s and '70s offered advice freely, and considered it appropriate and in some cases even necessary to do so, why do so many scientists feel so differently now?

There are no doubt many factors that bear on a fully adequate answer to this question, but the history recounted here suggests an important factor that is not often noted: that when it comes to the relation between science and policy, contemporary scientists have transformed necessity into virtue and political reality into epistemology. The interplay between scientific attitudes toward cultural engagement and the cultural attitude toward science—and between the demands of objectivity and loyalty—suggests that to a significant extent, what scientists think they should do depends on both the perception and the reality of what they can do, given the prevailing political and social context. Many of the scientists working on the issues we have studied here either faced or were aware of potential resistance to their findings from their governments, from industry, or in the broader culture, and this has shaped the structure of scientific assessments. In what follows we show how participants tried to keep their assessments safely on the science side of the science/policy border, how they embraced neutrality as a marker of objectivity, and how they pursued inclusiveness as a defense against alleged bias.

BORDER TROUBLES

Both NAPAP and early ozone assessments included both science and policy. Sometimes this was by design, as when scientists working on acid rain were asked to "identify actions to limit or ameliorate the harmful effects of acid precipitation" or when scientists assessing ozone depletion were asked to evaluate the urgency of the issue.[50] Scientists were criticized for this—by industry representatives, by government officials, and by later commentators—but they were responding to their charge.

In chapter 2 we saw how NAPAP scientists were accused of compromising their credibility by blurring the boundary between science and policy, thus delaying regulatory action. In chapter 3 we saw how scientists at the US National Research Council (NRC) assessing ozone depletion were criticized for wading into policy waters in their 1976 reports, despite their attempt to segregate science from policy by writing separate reports. The questions that the NRC scientists were asked to answer—which included both whether chlorofluorocarbons (CFCs) would destroy ozone and "the policy consequences of both our present knowledge and the knowledge we are likely to have in the future"—were partly scientific, but not entirely so.[51] To assess the threat and its policy consequences, scientists had to estimate the amount of ozone depletion that would occur over a given time frame and the extent of harm that would ensue. The policy and the science were necessarily intertwined.

Scientists were being asked to address a question that was at once scientific and political: How bad is this problem and therefore how rapidly do we need to address it? In essence, it was a question about the urgency of a proposed policy goal. The scientists agreed that CFCs could destroy ozone—and this was a bad thing—but the severity of the threat was not easy to judge. As technical experts, these scientists were in a good position to offer their judgments—but they were making judgments, not reporting scientific facts. It was not unreasonable that they were asked to offer these judgments. Their expertise placed them in a position to give informed answers. But the answers they provided could not be purely scientific. As Eisenhower had recognized, the terms "science" and "policy" are too blunt to capture the subtleties of the issues at stake.

The participants in the NRC assessments recognized the challenge that they faced, but as they were grappling with the questions placed in front of them, the relationship between scientific experts and the US federal government was also changing. The scientists did what they were asked to do, but by the time they had done it opponents were ready to challenge it. Industrial groups had already started to push back against the scientific evidence of ozone depletion, and so the NRC scientists tried to find a way to preempt the criticism that they anticipated while still answering the questions that had been put to them, by writing separate reports.

Consider another example, also from the ozone story (chapter 3). In 1975, scientists in the Climate Impacts Assessment Program concluded that the exhaust produced by a proposed SST fleet would pose a serious threat to the ozone layer. This was information that was policy-relevant but not prescriptive: the scientists assessed the potential impacts of an SST fleet but took no explicit position on whether the airplane should be built and the fleet deployed. However, assuming that one understood the role of the ozone layer in protecting life on earth, the conclusion was implicit. The bureaucrats in the sponsoring agency (the US Department of Transportation) distorted the message by writing an executive summary that deflected attention from the implicit conclusion. The summary focused on the effects of a small, near-term projected SST fleet (30 or so aircraft)—which were essentially negligible—and downplayed the possible effects that the scientists had addressed of a large, long-term projected fleet, which were not negligible at all. In addition, any potential adverse effects were cast in the executive summary as preventable through future unspecified and as yet undeveloped technology. Scientists objected, but the damage to their credibility had been done.

The crucial point here is that it mattered little whether the policy findings were implicit or explicit or whether or not there was a good-faith effort to police the science/policy border. Scientists had honored the science/policy border, but that did not protect their work or credibility from those who felt threatened by their conclusions.

These episodes suggest that as much as scientists strive to be fair, neutral, and objective and to demarcate scientific findings from policy recommendations, or even to avoid making recommendations at all, the intrinsically political character of assessments in the contexts in which they are produced and consumed makes it almost inevitable that there will be pushback. Individuals, groups, institutions, and economic and political actors will challenge the science and even the scientists when they feel that their interests are threatened. Pushback of this sort will cause some scientists to retreat from the contested borderland, to situate their work as deeply as possible in the technical domain, and to become even more scrupulous about avoiding any suggestion of policy recommendation.

This was the lesson that Bob Watson and Dan Albritton took from these experiences: that assessments should work harder to articulate a bright line between science and policy. Assessments should be policy-relevant but "never . . . prescriptive." Watson expressed pride in the formulation he developed with Dan Albritton, citing the example of the finding that "Unless there is a 100% elimination . . . of all long-lived chlorine- and bromine-containing compounds . . . the Antarctic ozone hole will be with us forever." This, he argues, was not prescriptive because it did not tell the governments what to do. "It was totally non-prescriptive," he insisted. But the implications of Watson's and Albritton's formulation are obvious. Unless one thinks that it is acceptable to allow the ozone hole to persist forever, and life on earth to perish, then it is clear that there should be a "100% elimination . . . of all long-lived chlorine- and bromine-containing compounds."[52]

To avoid prescription and the pushback they thought it could provoke, both NAPAP and ozone scientists moved into the mode of "scenario development"—outlining what-if (or what-if-not) options—that is now used extensively in climate assessments. But, as several of our informants have noted, his still implicates them in choices that are not purely scientific. When climate scientist Jonathan Shanklin argued in an interview for the benefits of letting politicians "choose from a menu" of policy options, his colleague Michael MacIntyre revised that to say that scientists should not present policy options but should say, "If you do this, then we think the range of possible [outcomes] is that."[53] Yet, whether it is a menu of options or a set of scenarios, scientists largely decide what is on the menu and which set of scenarios is reasonable and appropriate to analyze.

Watson's choices for policy makers and Shanklin's menu of policy options thus introduce an important tension. On the one hand, ozone assessors now generally agree that it is not their place to make explicit policy recommendations; the international ozone assessments since the Montreal

Protocol have adhered to this ideal. On the other hand, assessors also agree that the assessments should present a clear set of options, menu selections, or choices. Climate scientists face a similar situation: the options they present will inevitably reflect their preferences and orientations, at least to some degree, as well as their implicit recommendations. The IPCC presents a high-emissions scenario that is understood to represent "business as usual," but the context of the report makes it clear that choosing that option would be profoundly ill advised.

This is one reason assessments come under attack as politicized by those who think that doing nothing is acceptable and perhaps preferable. As already noted, the very existence of the assessment suggests there is a problem about which something should be done. The IPCC "business as usual" scenario (RCP8.5 in the *Fifth Assessment Report*) is presented not as a reasonable choice but as a means to demonstrate the adverse implications of continuing our current practices. In principle, business as usual is one of the options, but in practice there is an implicit message that it would be highly undesirable if not unconscionable. In the history of ozone assessment, it was always implicit that the goal was ozone protection; no ozone scientist ever publicly (or privately, so far as we can determine) suggested that the issue should be ignored.[54] Skeptics did not argue that destroying the ozone layer was acceptable; they argued that they were not persuaded by the scientific evidence that it would be destroyed.[55]

Assessors are not telling policy makers what choices to make, but they are deciding what choices to present and guiding policy makers to interpret those choices in certain ways. They are not presenting every option under the sun; they are presenting a set of options that seem reasonable to them. Scientists (and governments) routinely ignore options that others might consider reasonable. Prayer is one obvious example; grassroots organizing leading to rapid social and technological transformation is another. The conceptual virtue of scenarios is clear, but the strategy does not expunge judgment. The border between science and policy remains porous.

NEUTRALITY AND CREDIBILITY

Scientists and others are often concerned to distinguish science and policy because they interpret objectivity (at least in part) as policy neutrality and view objectivity as necessary for credibility. Edward Parson has criticized scientists involved in the NRC ozone reports for venturing into the policy domain, arguing that this "established a harmful model for scientific assessments" because the scientists were perceived as taking sides.[56] In our interviews, scientists made a similar point: that they should avoid making policy recommendations to protect their credibility. But scientists who raise concerns about this issue rarely offer any details or provide evidence of an assessment in which this sort of credibility loss occurred.[57] Rather, the equation of objectivity with policy neutrality seems to be an article of faith. As we saw in chapter 3, some ozone scientists feared that their assessments would be viewed as biased if Sherwood Rowland participated because of the public stance he had taken on the need to control CFCs. However, there seems to be little evidence that that fear was justified.

As noted in chapter 1, we can find examples from the nineteenth century in which scientists offered policy recommendations in assessments, with no evidence that they lost credibility among the government officials to whom their reports were addressed. We have also seen that many scientists in the 1950s and 1960s took strong policy positions on diverse matters such as the hydrogen bomb, the Vietnam War, the threat of urban air pollution, and the costs and benefits of widespread DDT use. Eisenhower, we have already noted, wanted his advisors to give him advice not just regarding the facts of a problem but also on what to do about it. John Kennedy spoke of Rachel Carson with respect and admiration.[58] Mikhail Gorbachev stated that his motivation to control nuclear weapons and avert a nuclear winter came from learning from scientists about the severity of the problem; we know of no evidence that Gorbachev doubted the evidence because some of the scientists responsible for it had called for arms control.[59]

Many scientists believe that Carl Sagan lost credibility because of his visibility as a sentinel of nuclear winter; this is cited by scientists as a reason why they should be cautious about stepping into the limelight.[60] But scientists (and perhaps some commentators) have conflated three issues. One is whether an individual scientist should be excluded from an assessment because he or she has taken a public position on policy. A second is whether assessments themselves should exclude policy recommendations in order to be perceived as objective and therefore credible. A third is the impact on the reputation of scientists within the scientific community when they take public positions on policy. While it is outside the scope of this study to address this last point, it seems that many scientists do frown upon colleagues who take public positions, and there is a risk that outspoken scientists may lose standing among those colleagues.[61] But this is a different question from whether they, or their reports, lose credibility with the agencies or governments who have commissioned the work.

Many parties had reason to want to delay regulatory action on ozone, just as many have reason to delay action on climate, and it is likely they would have found reasons to justify their position irrespective of what scientists had done. The IPCC takes great care to avoid policy recommendations, but this has protected neither it, as an institution, nor its scientists from attack. Indeed, over the lifetime of the IPCC, several of its members have been publicly excoriated, most notably US scientists Benjamin Santer and Michael Mann, and the UK scientist Phil Jones. Santer was the convening lead author of the 1995 IPCC report chapter that stated that the human fingerprint in the climate system was "discernible"; Mann and Jones had helped to develop important historical temperature reconstructions that demonstrated that recent observed temperature increases were outside of the envelope of the natural variability of the past millennia.[62] All three have been the target of public vilification; Jones was the target of the stolen email affair that came to be known as "Climategate."[63] None has ever been shown to have committed scientific misconduct of any kind.[64]

After his work was challenged, Santer took pains to set the record straight with his colleagues—to ensure that he had not lost credibility with them. He did this because he understood that the future of his career depended upon it. But if either Santer or the IPCC lost credibility in this incident, it was not because they had strayed into policy space; the same is true of Mann and Jones. None of these men had advocated climate policy.

Nevertheless, a number of journalists and social critics blamed them, suggesting that they had brought these attacks upon themselves by being arrogant and advocating policy. Consider this comment by John Broder in the *New York Times* in the wake of the stolen email incident: "Climate scientists have been shaken by the criticism and are beginning to look for ways to recover their reputation. They are learning a little humility and trying to make sure they avoid crossing a line into policy advocacy."[65] But Broder offered no evidence to support this claim—nor could he, because Phil Jones had never advocated climate policy, nor had Santer or Mann.[66] If Broder were right—that avoiding policy questions helps to protect scientists from criticism—then we might expect to see fewer disputes about climate change and its assessment than we did for acid rain and ozone. But that is not the case. We have discussed how IPCC has made extensive efforts to be "policy-relevant but not policy-prescriptive," but that has not protected it from politically charged critique. There are many reasons why climate science is contested, but in the context under discussion here, it makes more sense to see the retreat from policy—both on the part of individuals and in formalized assessments—as a response to political contestation rather than seeing political contestation as response to an insufficient retreat from policy. Scientists have not been attacked because they have failed to separate themselves sufficiently from policy; they have been attacked because it is impossible to separate their work entirely from social and political implications. As scientific findings about climate change have become increasingly alarming, accusations of "alarmism" have increased accordingly. But this is not because scientists have strayed into the policy domain.

The evidence presented here suggests that when scientists feel vulnerable, they retreat from policy recommendations. Scientists in the late twentieth century found themselves in a weaker position than their counterparts in midcentury, as the cultural preeminence of science declined, the postwar consensus frayed, and political polarization increased. As a result of these (and other) factors, opposition to scientific findings that challenged the status quo became increasingly well organized and well funded.[67] Since the 1980s, scientists have faced pushback not just in the arena of climate change but also in domains related to endocrine-disrupting chemicals, lead poisoning, tobacco, the safety of vaccinations, and other matters.[68] In this environment, scientists have looked to the traditional scientific norms of objectivity and value-neutrality to demonstrate their trustworthiness. Because their colleagues share these values, the choice is ratified by them, and scientists conclude that it is the right choice. This tells us much about the values of scientists, but it does not answer the question of what scientists should do in assessments to address matters of concern while protecting themselves from politically motivated attack.

INCLUSIVENESS AND THE BALANCING OF BIASES

Another strategy scientists have taken to protect themselves from accusations of bias has been to make their assessments more inclusive and explicitly international. In particular, by making their ozone assessments international, scientists tried to address the complaint that they were biased in favor of their own governments' views. They also began to expand the size of their assessments to include as many relevant experts as possible. This approach is implemented at the IPCC today, in which inclusivity is a guiding principle. Authorship must include men and women from many countries, and the IPCC attempts to reduce the historical dominance of scientists from the United States or Western Europe by actively seeking scientists from elsewhere. In the IPCC, it is viewed as important that the author team be intellectually inclusive; to the extent possible, anyone who has significant expertise should be included in the process, if not as a lead author then at least as a contributing author or a peer reviewer. Anticipated accusations of bias are thus preempted through inclusionary processes: the intellectual presumption is that so long as diverse voices are heard, no one particular bias can prevail. We may call this a "balance-of-bias" approach—the assumption being that although bias cannot be eliminated, it can be compensated for balanced by opposing biases.[69]

The expansion of the IPCC to be as inclusive as possible can be viewed as reflecting a vision of objectivity as intersubjective agreement within a diverse community rather than as a characteristic of an individual.[70] This underscores the argument introduced in chapter 1 that the modern assessment reflects a significant change in scientists' conceptions of expert knowledge. In early modern science, the reliability of the knowledge produced was assumed to arise from the stature and reliability of the individual or individuals involved. As Steven Shapin and others have emphasized, early modern traditions placed the source of epistemic credibility in the virtues of the individual scientist. This view persisted into the mid-twentieth century, when we still find small groups of "wise men" called upon to offer up expertise on diverse subjects, in some cases ranging far from their disciplinary expertise.[71] The intellectual presumption was that if good men were chosen, good answers would follow.

The modern assessment both reflects and creates a different epistemological standard, one that suggests that no matter how "good" any particular expert may be, his or her views alone offer an insufficient basis for reliable knowledge. Objectivity in assessments is displayed not by finding the right (unbiased) individuals but by finding a capacious and comprehensive mix of differently biased ones. Bias is viewed by scientists as a form of error that may be compensated for—if not entirely canceled—by opposing error. This represents a significant epistemic shift from locating the source of scientific objectivity and reliability in the individual to locating it in an institutional process. The balance-of-bias approach gives the scientific community an argument with which to respond to accusations of bias (whether it produces an epistemically more robust result is another matter).

Despite scientists' commitment to locating objectivity in the assessment process, important cultural strands in the United States and Europe cling to the older, more individualistic model; this opens assessments to criticism even when they are inclusive and diverse. As already noted, opponents of action on climate change have embarked on significant efforts to discredit particular individuals whose work has played a major role in IPCC conclusions. Although the attackers may present themselves as "skeptics" offering scientific dissent, these attacks are largely driven by political considerations.[72] So we should not expect these "skeptics" to be persuaded by the various techniques that the IPCC has embraced to try to ensure a fair and objective product. However, we could expect that if the IPCC techniques were culturally persuasive, then attacks on individual scientists would have little cultural resonance. If journalists or other observers accepted the "balance of bias" model, it might not matter to them if one or even a handful of scientists were shown to be foolish, mistaken, or even venal. Yet press coverage and public opinion polls clearly show that it does matter.[73] Assessments remain vulnerable to public attack in part because the model of objectivity applied is not one that has been generally accepted by the public.

THE SCIENTIST AS SENTINEL

Even in the 1950s, when the credibility of scientists seemed to be at its peak, questions were asked about how much deference to give to scientists. Niels Bohr was criticized not only by government officials who suspected his motives but also by civilian commentators who questioned his authority to expound on matters of international diplomacy. Was Bohr not speaking out of turn when he attempted to tell world leaders how they should pursue their affairs? Arms control is not, after all, a scientific matter; it is a social and political one. Was it not ironic, even hypocritical, for the scientists who had made weapons of mass destruction to instruct the world on the necessity of peace?

These were reasonable questions in 1950, and they remain reasonable today. Should scientists speak beyond the domain of their technical expertise? If they do, what obligations do they incur? Certainly, scientists have the same right as ordinary citizens to engage in public discourse, but do they have an additional obligation to alert the world to threats, challenges, and opportunities of which, by virtue of their scientific expertise, they are especially or even uniquely aware? Conversely, do they have a special obligation not to exploit their position of intellectual authority and social privilege to advocate for particular policies?

One way to approach these questions is through the following thought experiment. Imagine that Sherwood Rowland and his colleagues had not publicized their research demonstrating that CFCs had the potential to destroy stratospheric ozone. Imagine, instead, that they had published it only as articles in peer-reviewed journals and that, like most scientific work, it had been ignored outside their expert community. Now imagine that 30, 40, or 50 years later, dermatologists and oncologists began to notice a significant but unexplained increase in rates of skin cancer. Epidemiologists analyzed the available data and concluded that there was an epidemic of skin cancers around the globe, and it was especially severe in Australia and southern Chile, and among white South Africans. Meanwhile, plant pathologists and horticulturalists noticed increased UV damage in agricultural crops; veterinarians noted increased rates of cataracts in farm animals. Scientists would have begun to search for an explanation for this strange association of human, animal, and plant pathology. In time, someone would have come across Rowland's work, connected the dots, and understood what was happening. Programs would then have been put in place quickly to measure stratospheric ozone, which would have demonstrated that the ozone layer had been severely depleted. But by that point, it would have been too late to avoid grave damage.

This thought experiment makes it clear that society owes a great debt to Rowland and his colleagues who acted as sentinels on the ozone issue. Society needs scientists to be sentinels on issues like ozone or acid rain or climate change (or emerging epidemics) because laypeople are not in a position to appreciate these sorts of threats or in some cases even to know that they exist. Scientists, by virtue of their specialized knowledge, may be in a unique position to discern things that are not apparent to others. In the 1970s, ozone scientists (alone) understood the threat that ozone depletion presented. We needed them to be sentinels. We needed them to be discerning experts.

However, it is one thing to alert society to a problem, another to tell society what to do about it, and a third to instruct society on how to do it. This suggests that a possible approach to analyzing scientists' role is to differentiate problems from policies and policies from instruments. Rowland's colleagues Bob Watson and Dan Albritton concluded that as physical scientists they should not make policy recommendations: it was acceptable to identify the problem and present their findings, but not to recommend a solution. The realm of solutions moved them into policy, and that was the realm of policy makers (and, of course, other types of expertise, including social scientific and humanistic).

But, as we have seen, the boundary between findings and recommendations is not always clear or even determinate. The questions, Do CFCs need to be phased out? and If so, how soon? were policy questions in the sense that they were about what to do. (Similar questions arose about acid rain and climate change.) But they were also scientific questions because it took scientific expertise to determine how much CFCs needed to be reduced to protect ozone, just as it takes scientific expertise to answer the question, How much do greenhouse gases need to be reduced to keep global temperature change below 2°C?

The questions, Should we ban CFCs? and Should we put a price on carbon? are different still. These are questions about instruments: about how best to achieve a goal, assuming we agree on it. For Rowland, once one understood that CFCs were destroying ozone—and taking it as obvious that one did not want life on earth to be destroyed—it was a logical consequence that CFCs had to be controlled. But to decide how to achieve that control required different sorts of expertise: in economics, politics, law, or perhaps moral theory. We could call this second sense of policy the "how to do it": with taxes, treaties, emission-trading regimes, or other policy instruments.

Rowland's position implicated him in a value premise: the value of life on earth as we know it. If one wanted to protect life on earth, then it was necessary to prevent ozone depletion. For Rowland, the value of life was so obvious as not to need stating, so the implication that ozone needed to be protected was equally obvious—and none of his colleagues ever argued otherwise. At least in the early years, he did not advocate a specific policy instrument; he simply spoke strongly in public about the urgency of the problem. After 1986 or so, Rowland did call for a ban on CFCs, but even before that point some of his colleagues felt that he went too far—that he was too outspoken in calling for action and therefore should not be asked to serve on ozone assessments. How did they make that judgment? Why was it acceptable to imply that ozone-depleting chemicals needed to be controlled but unacceptable to say so explicitly?[74]

We have suggested that part of the answer lies in increasing external pressure on scientists in the late twentieth century. Our case studies show that leading scientists concluded that one way to protect themselves from criticism would be to stay away from policy (and therefore, implicitly, values) and to develop rhetorical and epistemic strategies that articulated and reinforced a border that they promised not to cross.[75] This shift involves rejecting the role of a "total intellectual" who has "domain-general" competence in favor of that of a "specific expert" whose competence is "domain-specific."[76]

As an example of scientists marking the limits of their authority in this way, consider the following case. When interviewed by the *New York Times* on the occasion of the release of the IPCC *Fourth Assessment Report*, cochair

Susan Solomon reiterated the IPCC conclusion that "warming was unequivocal." But when asked what should be done about it, Solomon replied, "It's not my role to try to communicate what should be done."[77] Clearly there is merit in this position: outside their domains of expertise, scientists may be quite ignorant, often knowing little more than laypeople and sometimes knowing less as a consequence of their long years of specialized training and acutely focused work—what Pierre Bourdieu called their "militant craftsmanship."[78] When scientists decline to comment on the policy dimensions of global warming, they are acknowledging the limits of their craft, as well as preempting the claim that their science is biased by their political preferences.

Yet our discussion should also make clear that many pressing issues such as the challenge of climate change cannot be solved by specific expertise alone. Diverse actors from Dwight Eisenhower to Bob Watson have noted that policy choices involve a good deal more than technical considerations and, as we have suggested, scientific assessments are embedded in complex webs of societal values. We have also seen that scientists in earlier generations gave advice on variegated matters, and often it was good advice (Niels Bohr was right about the arms race; Rowland was right about CFCs). At least one scientist of the generation who freely gave policy advice on diverse matters felt that Solomon was too reticent: former Caltech president Marvin "Murph" Goldberger, a member of PSAC during the 1960s and cofounder of the Jasons, felt that Solomon had missed an important opportunity.[79]

Societies need advice, and not just of the narrowly technical sort. One function of assessments is to bring together diverse expertise in order to avoid the pitfalls of overspecialization and to generate the integrating visions needed to solve complex problems. However, while scientific assessments bring together different areas of science, it is often only natural science. Expertise in communication, ethics, politics, theology, or morality is generally not included.[80] This may be changing as the IPCC includes more experts in social vulnerability in Working Group II (on impacts and adaptation) and ethics in Working Group III (on mitigation of climate change). It is significant that the work of both groups is framed in the context of sustainable development—a concept that has a technical dimension but is primarily social, economic, and moral.[81]

WORKING FROM PROXIMATE EXPERTISE

Let us return for a moment to our thought experiment in which Sherwood Rowland and his colleagues did not speak out publicly but only published their findings in peer-reviewed journals. While counterfactual, this idea is not fantastic: it is essentially what did occur with asbestos and tobacco;[82] it could easily have been the case with CFCs. The argument is not that the world needed Rowland, as an individual, to sound the alert but that we did need someone in that epistemological community to do it.

Chemists or atmospheric scientists had to be the ones to alert the world to the idea that certain chemicals could destroy the protective ozone layer because they were the ones who had this information at their disposal. Once it was made known, then oncologists and dermatologists and many others could comment on the expected adverse impacts of increased UV exposure—and many did. Indeed, once ozone depletion was shown to be under way, then various forms of biological expertise became extremely important to understanding why it mattered. But it was chemists and atmospheric scientists who first realized that ozone depletion might occur and therefore had to be the ones to sound the alarm. By virtue of their epistemic proximity to the problem, they were in a position to discern it. Thus we may conclude that the world needed ozone scientists to speak up about ozone depletion because they were the proximate experts. In speaking up, these experts did not step outside their disciplines, but they did step beyond the confines of their disciplinary norms, and it was this, perhaps, that made some colleagues uncomfortable.

We might nevertheless argue that it is one thing to say, "CFCs can destroy the ozone layer that protects life on earth from damaging UV light" (a statement of scientific fact) and another to say, "We ought to take steps to control CFCs" (a policy recommendation). Moreover, the second statement is a consequence of the scientific information only when it is supplemented with the (in this case noncontroversial) premise that we want life on earth to continue. What Rowland and others were arguing, in the wake of their discovery that CFCs destroy ozone and this threatens life on earth, was

If we want life on Earth to continue, then we must control CFCs.
We want life on earth to continue.
Therefore we must control CFCs.

The NRC committees and World Meteorological Organization ozone assessors then went further to suggest: Given the rate at which CFCs are destroying ozone, we need to reduce CFCs by *x* amount in *y* time if we want to stop ozone loss from getting worse.

Clearly, this moves us further in the policy direction, yet it is hard to credit the criticism that this was inappropriate, because using scientific expertise to understand causes and rates is something scientists do every day. However, suppose that we say:

> Therefore, we should implement an international treaty to phase out the use of CFCs by Annex I countries by 1996.

This is clearly a policy recommendation that takes us beyond science or the proximate expertise of an ozone scientist.

The point here is not that the policy recommendation violates some ideal of value neutrality; values are present all the way down. Rather, it is a point about proximate expertise. By virtue of their expertise, ozone scientists are in a position to go beyond the initial discovery that CFCs deplete ozone and to claim that we should control CFCs, to calculate the consequences of various emissions scenarios, and to suggest the rate of CFC control that would be needed to protect the ozone layer. But to evaluate particular policy instruments requires different forms of expertise.

CONCLUSION

Despite decades of social scientific scholarship demonstrating the complex ways in which science and policy—facts and values—are intertwined, scientists still strive to keep them distinct.[83] We suggest that the injunction to keep facts and values distinct can be viewed as a regulative ideal, one that can serve useful purposes, including to remind scientists of the limits of their role and expertise, but should not be viewed as anything like a categorical imperative.[84]

Scientists involved in assessments generally recognize many of the complexities involved, but in a variety of public contexts they sometimes fail to recognize when their expertise is proximate to certain kinds of questions and distant from others. It is our view that scientists should not be reticent about interventions where their expertise is proximate, but they should be reticent regarding interventions in which their expertise is not.

CHAPTER SIX

What Assessments Do

INTRODUCTION

Scientists, representatives of governments and international organizations, and diplomats put enormous effort into assessments. But what precisely do assessments do? In general, scientists involved in assessments distinguish them from research. Clearly, assessment involves reviewing, evaluating, and judging knowledge, but how should the product be characterized? Do assessments produce new scientific insights and understanding? If they produce new knowledge, what kind of knowledge is it?

The US National Acid Precipitation Assessment Program (NAPAP) explicitly funded and oversaw research that created new scientific knowledge, but in this regard it seems to be exceptional. Many of those whom we interviewed in the course of our studies drew a sharp distinction between assessing an existing body of knowledge and creating new knowledge. Many of our informants said that the assessments on which they worked produced no new knowledge. The Intergovernmental Panel on Climate Change (IPCC) explicitly states that "it does not conduct any research nor does it monitor climate related data or parameters."[1] If the fruits of research can be equated with "new knowledge," this organizational statement would seem to close the matter. As we shall see, it is not so simple.

Assessments clearly do more than simply summarize existing scientific data. In our case studies there are instances in which assessors acted much as research scientists do, developing new findings that were sufficiently novel to be published as research (e.g., the 1983 National Research Council acid report discussed below). IPCC assessors, working in an institution that explicitly eschews doing research, nevertheless sometimes have done so, particularly when trying to assess highly uncertain science.

One of the points of an assessment is to improve our epistemological position. This raises the question of whether new knowledge is produced in the process, since that is one obvious way of improving our epistemological position. Clearly, assessments can clarify what is known, narrow uncertainty, increase our confidence in prior beliefs, lead to new hypotheses and explanations, and help integrate and update prior beliefs and models. But assessments can also widen uncertainty or decrease confidence in prior beliefs. The output of assessments can be characterized along a spectrum ranging from, at one extreme, new research that produces new knowledge or destabilizes old knowledge to, at the other extreme, reviewing and characterizing existing research. In any case, the output of assessments is supposed to be more accessible and useful to policy makers than the knowledge that goes into it. But the very newness of all of the activities that occur within an assessment raises questions about the reliability of output that has not yet stood the test of time.

The reluctance to produce new knowledge may be well founded, because new knowledge is less well tested, and therefore may be less reliable, than established knowledge. For something to count as established scientific knowledge, it must be certified by the standards of expert communities for vetting knowledge claims, such as peer review. If an assessment produces new knowledge, it will have bypassed the standard process. Assessments generally establish their own processes to review drafts of the reports they produce, and these sometimes have more stages of review and involve far more reviewers than the typical review performed by a professional journal. This is certainly the case for IPCC and for the international ozone assessments sponsored by the World Meteorological Organization (WMO), the US National Aeronautics and Space Administration, and several other agencies. However, assessment review generally lacks an independent arbiter with final authority who has little or no stake in the outcome, such as a journal editor. Assessment authors have far more leverage in deciding a contested issue than do paper authors, and they can be confident that in some form the chapter will be published.[2]

In addition, if assessors produce new knowledge as part of the assessment process, without independent vetting, it may undermine their neutrality as assessors. Individuals in an assessment may try to promote their own work, but the assumption is that this self-interest is balanced by the competing interests and perspectives of others in the group. However, if the group itself creates new knowledge, then that element of balancing interests may be missing. Manuscripts submitted to journals and then subjected to evaluation in an assessment are scrutinized in two independent processes: one by the journal reviewers and the other by the assessors. At best, new knowledge developed by assessment authors is vetted in only one process (which, like a journal review, may have multiple rounds), when the assessment itself is reviewed.

These considerations are especially important because the knowledge that an assessment produces is passed on to policy makers. The presumption is that any resulting policies are based on well-vetted underlying science. However, if the assessors have produced novel claims as part of the assessment, then that presumption is questionable. In addition, decisions made in the assessment process can reframe a scientific problem and introduce conventions that have substantive effects on how a policy problem is understood and what counts as a solution. Examples include designating global mean surface temperature as the measure of climate change and the choice of a particular pH value as the criterion for lake acidification.

Both scientists and decision makers see uncertainty as important to risk management, and the characterization of uncertainty is a major task of most assessments. However, "uncertainty" is dense with meaning and associated with diverse views about the relationship between science and policy.[3] Assertions of uncertainty imply both the possibility of certainty and presumed paths from one to the other. Claims of uncertainty reflect and establish epistemological order and can suggest the need for both particular research programs and particular policy approaches. Scientific uncertainty mediates between the worlds of scientific knowledge and public policy formation by raising the question of how uncertainty and risk are to be managed—for scientific, social, and political purposes.

The three groups of assessments we have studied focused heavily on integrating knowledge, while reaching beyond the published peer-reviewed literature in order to characterize uncertainty in qualitative and quantitative terms. If knowledge is understood to include relating different types of knowledge and elaborating uncertainties, then it seems that assessments normally result in new knowledge.

NAPAP: CHOICES AND CONVENTIONS

Part of NAPAP's mission was to fund a research program aimed at producing new knowledge, including basic scientific information about the sources of acid precipitation and its impacts. NAPAP did what it was asked. One example was the comprehensive study of the chemical characteristics of lakes in the eastern United States. NAPAP also situated the data in a policy-relevant context by using it to model potential future change in lake water chemistry as a result of acid deposition.

In terms of objective and intended audience, the executive summary of the 1987 Interim Assessment (as opposed to summaries of its research programs) is akin to the assessment products (executive summaries and summaries for policy makers) from WMO and IPCC on ozone and climate change, respectively. What sort of knowledge, or epistemological advance, did the process of developing the NAPAP executive summary produce? Addressing this question also provides some insight into the knowledge produced by the WMO and IPCC assessment processes.

Consider NAPAP's assessment of the extent of acidic surface waters (e.g., lakes and streams). As we saw in chapter 2, the NAPAP executive summary defined an acidic lake as one in which the pH is less than 5, thus excluding as acidic some lakes in which there was evidence of biological damage but whose pH was 5.5 or 6. Since the scientific work was already under way, the choice of pH 5 in the executive summary did not interfere with NAPAP knowledge production about lakes with higher pH, but it effectively implied that damage occurring in lakes with pH above 5.0 was inconsequential. This resulted in minimizing the problem of acid precipitation and may have influenced future research directions. It also exacerbated the gap between the US and Canadian perspectives, since the Canadians (and Norwegians), guided by the biological evidence, used a reference value of pH 6. By defining lake acidification only in terms of a chemical characteristic and excluding biological, geophysical, and human interactions, the executive summary effectively "subtracted" knowledge that was available in the primary literature.

Even from a narrowly chemical perspective, the adoption of pH 5 as the criterion of acidification was an innovation. High school textbooks typically characterize acidity as involving a pH of less than 7. NAPAP's choice was motivated by the fact that scientists had not found a great deal of damage in waters that were only slightly acidic. What appeared to be a purely chemical criterion was in fact based on a view about damages.

The conventions of a discipline, or those adopted in an assessment, help to determine what questions are asked and what counts as an answer. Where policy and science interact, as they do in the context of assessments, conventions and other boundaries take on critical importance. By choosing a convention that defined the boundaries of the acid rain problem in chemical terms, NAPAP influenced what policy makers would focus on and therefore what scientific information they might have sought.

Another issue that involves choice and convention is the scale of analysis. Acid rain assessors confronted the question of whether the relation of sulfur dioxide emissions to the concentration of acid in precipitation was linear or nonlinear (i.e., how constant is the relation between the inputs [emissions] and the outputs [concentration of acids]). At a sufficiently small geographic scale of observation (or modeling) or a sufficiently high level of precision, this relationship is inherently nonlinear (as are in any relationships in nature).

There are two levels of scientific interest: the process level, which considers molecular-scale interactions (e.g., chemical reactions of molecules of the acidic pollutant sulfur dioxide in the atmosphere), and the ecological level, which considers whole systems such as lakes, watersheds, or entire forests and emissions from groups of sources rather than any single smokestack. At the latter scale, one cares less about what might happen in any given day or week and more about long-term averages over seasons, years, or decades. Nonlinearities that are important to the disposition of any cluster of molecules emitted by any one smokestack tend to average out to linear relations connecting groups of smokestacks to impacts on extended areas occurring over years. Generally, the latter scale matters much more than the former one to ecologists, water-quality experts, and policy makers.

The National Research Council (NRC)'s 1983 assessment of the linearity question emphasized this large-scale aspect for the first time.[4] It did so by drawing on a very limited amount of existing literature and also developing its own method of assessing the problem (based on the relative amounts of sulfate and nitrate pollution in precipitation). This was new knowledge for both scientists and policy makers. The NRC panel reframed the question by changing the scale of concern from the micro to the macro and then providing a plausible answer to the linearity question. As a side benefit, this approach yielded insights into micro-level processes of interest mainly to scientists. This shows that shifting the scale of interest can be central to producing new knowledge. The decision to reframe and restrict the boundaries of the scientific problem to the larger scale was essentially a matter of choice or convention—and when viewed at that scale the relationship between emissions and acidity was effectively linear, a result that was potentially of great importance to policy makers.

Sometimes assessors confront the problem that the peer-reviewed literature, in either its content or its framing, is inadequate to address issues that are important in policy decisions. Once the NRC panel reframed the emissions/acidity relation as a macro-level question, they were confronted with the fact that there was insufficient literature to answer it. The panel proceeded to perform some research—to create new knowledge—in a very limited amount of time. As we saw in chapter 4, even the IPCC, with its explicit goal of not performing research, occasionally makes the decision to perform new research as well.

NAPAP assessors chose a different course, launching a multiyear modeling project and creating the Regional Acid Deposition Model (RADM) to explore the cumulative effect of the various micro-scale processes. In effect, NAPAP either resisted the redefinition of the scale of the problem or sought to answer the macro-scale question using a reductionist approach (i.e., by letting the answer emerge from the integration of micro-level processes that could be revealed by modeling). NAPAP could have developed new knowledge on the linearity question in the way the NRC panel did, either in its summary or in the main body of the assessment, but it chose not to do so.

The NRC and NAPAP were working under different institutional conditions and constraints. The small size and tight deadline of the NRC panel may have encouraged its members to offer an answer to the linearity question. Scientists at government agencies experience different pressures and incentives than do scientists on NRC panels. Indeed, the departments involved in NAPAP, particularly the Department of Energy, heavily and publicly criticized the NRC approach. Chapter 2 explains the various reasons for this outcome: questions of politics and agency turf and priorities were clearly at work. In contrast, the NRC panel had six philanthropists as sponsors, and institutional priorities beyond those of the NRC likely played a much smaller role for the panel members than within NAPAP.

In summary, the characterizations of both acidity and linearity helped to determine the answers to key policy-relevant questions, the tools used to determine the answers, and the degree of their utility to policy makers. NAPAP's definition of acidity restricted the number of surface waters of interest and transformed the problem, reducing its scope for both scientists and policy makers. By excluding a large number of lakes from consideration, it reduced the opportunities for learning about those lakes. Whether this change was an epistemological improvement is debatable, but it definitely affected the level of certainty about the problem by excluding the most uncertain part (i.e., impacts for pH > 5).

Had the NRC characterization of linearity been fully accepted by either scientists or policy makers, it would have constituted a very large step in reducing uncertainty (and creating new knowledge). Instead, its effect on uncertainty was limited because NAPAP persisted in keeping the linearity question open, pursuing RADM and criticizing the NRC's macro-scale approach. As a consequence, uncertainty and even confusion persisted over the linearity issue in the scientific world as well as the policy world. When the Clean Air Act was amended in 1990 to address acid rain, it effectively ended policy interest in this question, long before RADM could be completed.

OZONE ASSESSMENTS: KNOWLEDGE, UNCERTAINTY, AND IGNORANCE

The period 1985–1995, beginning with the reported observation of the ozone hole, was characterized by extremely rapid development of scientific understanding of ozone depletion and an unusually intimate peacetime interaction between evolving science and international policy. The Montreal Protocol was signed in 1987 and was followed by a series of amendments that progressively strengthened and broadened the protocol's restrictions on production of ozone-depleting substances. Scientific findings were a key input to the political negotiations.

The Montreal Protocol had not eliminated the substances responsible for the problem, and the key policy issue after 1987 was how to phase in deeper reductions than those required by the protocol. As for acid rain, scientists sought to understand the relationship between the amount of pollutants in the atmosphere and the environmental damage that would ensue. In other words, if emissions were cut by a given amount, what benefits would accrue? How much should emissions be reduced and on what schedule in order to induce recovery of the ozone layer?

The usual means for projecting the consequences of a policy intervention in such cases was to develop a "process-based" or mechanistic model based on the physics and chemistry that approximated the response of the atmosphere to emissions of the chemicals causing the problem. But the appearance of the ozone hole, which no model predicted, showed the inadequacy of the understandings on which the models were based. If assessors were to narrow the uncertainties and provide credible answers, they needed to transcend the limits of the available models. Unencumbered by formal limitations on what they were permitted to do, the assessors (in the 1990 and 1992 international WMO assessments)[5] broke new ground by applying surrogate measures of future ozone depletion: chlorine loading potential and equivalent effective stratospheric chlorine. These metrics had been developed to allow comparison of the depleting effects of the various chemicals involved, not for projecting the effects of future emissions—the purpose for which they were now used. Assessors thus filled a critical gap for both scientists and policy makers until improved process-based models became available toward the end of the 1990s.

In contrast, authors of the 1986 WMO-sponsored ozone assessment[6] shied away from pursuing the consequences of an identified uncertainty: the role of heterogeneous chemical processes in depleting ozone. As discussed in chapter 3, it had been hypothesized by some atmospheric scientists that chemical reactions that occur on small droplets floating in stratospheric air ("aerosols") could play an important role in ozone depletion. If this were true, then the potential for severe ozone depletion could be much greater than generally thought. However, the experiments needed to test this idea were extremely difficult to perform. In addition, much of the community assumed that the stratosphere had too few aerosol particles for these reactions to matter, and any reactions that did occur would be too slow to be significant. The reasoning based on reaction rates was sound given what was then known but turned out to be incorrect. The argument based on the absence of significant amounts of aerosol was also incorrect, but for a different reason: only a small number of atmospheric scientists familiar with the polar stratosphere knew that the cirrus-like polar stratospheric clouds observed there in springtime were in part constituted by such aerosol. Most of the atmospheric science community, focused on the midlatitudes above populated regions where observations were much easier, was ignorant of this. Thus earlier assessments had judged heterogeneous reactions to be insignificant. To the extent that assessments influence research plans, these earlier assessments took heterogeneous reactions off the agenda and contributed to a substantial underestimation of how serious the problem of ozone depletion was becoming.

The 1986 ozone assessment, at the time the most comprehensive overview of the problem, mentioned the aerosol issue as an uncertainty but dismissed its importance. Rather than take the now-routine risk-based approach and assign aerosol processes a finite (even if small) probability of occurrence, the assessors essentially assigned the possibility a zero likelihood. Alternatively, the assessors might have taken the approach adopted in the IPCC's *Fourth Assessment Report* (*AR4*) when projecting the ice dynamic contribution to sea level rise: declare ignorance and assert that the resulting projections could not be regarded as representing the outer limit of the possible (i.e., the upper bound).

Unlike the NAPAP assessors, the 1986 WMO assessors operated under no strong institutional constraints. WMO was not a strong organization, the Vienna Convention (out of which flowed the Montreal Protocol) was weak, and governments were not yet fully focused on the problem. The assessment scientists, mostly from the United States and United Kingdom, had a fairly strong culture of independence, even those employed by government laboratories and agencies. Like the 1983 NRC panelists dealing with the linearity issue relating to acid precipitation, the ozone assessors were not particularly bound by institutional constraints or by political demands for a timely response: no regulation of ozone-depleting substances was imminent (which was also true for the 1983 NRC acid precipitation assessors). Yet, unlike the NRC panel members who actively explored new approaches to the linearity issue, ozone assessors set aside the possibility of heterogeneous chemistry.

Whether this was good judgment given what was known or "erring on the side of least drama"[7] remains unclear. But as we saw in chapter 3, grasping the importance of heterogeneous chemistry ultimately required scientists to break out of disciplinary silos—experts in the polar atmosphere and experts in atmospheric chemistry needed to collaborate unusually closely. While assessments can provide opportunities for such collaboration and the potential for new knowledge to emerge, it is difficult to achieve and in the ozone case was only precipitated by the discovery of an extraordinary and frightening phenomenon, the ozone hole.

IPCC AND EXPERT JUDGMENT: SEA LEVEL AND *AR4*

We now turn to what proved to be the most fraught case of developing, or avoiding the development of, new knowledge. As discussed in chapter 4, authors of the IPCC's *Fourth Assessment Report* faced a quandary. In previous assessments, the IPCC had presented projections of sea level rise in the same manner as those for temperature, based on year-by-year output of models that represent the various processes contributing to warming and sea level rise. However, for *AR4*, IPCC stopped short of a complete sea level assessment because one of the three key processes contributing to future sea level rise, the transformation of parts of the Greenland and Antarctic Ice Sheets to ocean water, could not be reliably represented by any available model.

As described in chapter 4, ice sheets can both melt and flow into the sea (by forming icebergs), both effects changing land ice into sea water and raising sea level. The melting process is well understood, but the flow process, referred to as "dynamical," is not. Existing models failed to reproduce the current effect of dynamics. Furthermore, observations that became available just before and during the Fourth Assessment convinced the authors that the dynamical processes might in the future rival or exceed the other contributors to sea level rise. However, without a trustworthy model, the authors felt they had no basis to make a projection that attempted to estimate how the dynamic component might change in the future. Instead of doing as ozone assessors did when they redeployed the chlorine loading potential and equivalent effective stratospheric chlorine metrics for the purposes of projection, *AR4* authors published projections that included only the current dynamical component, estimated from observations (rather than models), and made no attempt to estimate (with the exception discussed below) how much this factor might increase as the world warmed.

Our objective is not to judge the wisdom of that decision or to revisit the controversy that ensued but rather to understand its epistemological consequences. The primary properties of the climate projected by the IPCC are global temperature change and mean sea level rise, and the decision on how to represent sea level rise was certainly not taken lightly (see chapter 4). Both institutional factors particular to IPCC and cultural factors in the earth sciences as currently practiced contributed to the outcome. Here we focus on those factors pertinent to the "new knowledge" question.

As we noted above, IPCC's "no research" rule is sometimes broken, and implicit understanding of this seems to be widespread within the IPCC community. This issue came to the fore in the *AR4* ice sheet assessment.

The *AR4* authors concluded that because they had no basis in modeling on which to offer an estimate of any increased dynamical contribution to sea level rise, they should not offer one at all. However, reviewers were dissatisfied and raised the question of why the authors did not attempt to estimate the dynamical component. In response, the authors sought a compromise: very near the end of drafting, a small correction was added, discussed in the text almost in passing, which provided one means to estimate the change in the dynamic contribution.

The "base case" assumption represented in the main sea level rise projections amounted to an assumption that dynamics would not change over the century. However, the dynamic term had already changed (and the authors believed it was likely to change further in the future) so the authors could justifiably propose various magnitudes of change for this term and test the effect of each on projected sea level rise—a method called sensitivity analysis. After lengthy consideration, the authors chose one way to estimate dynamics, assuming that the contribution would increase in proportion to temperature. They then pointed out, correctly, that this was one of many plausible assumptions. However, rather than present results for some of the others, which would be the usual procedure, they merely stated that there was nothing special about the chosen assumption and left it at that. This was a choice for which they were later criticized by many scientists outside the

IPCC. In offering up a half-hearted attempt at sensitivity analysis, to which they were reluctant to grant too much credence, the authors produced something that was new but clearly incomplete and perhaps misleading.

One clear case of the IPCC performing research is an entire report, the *IPCC Special Report on Emissions Scenarios* (*SRES*).[8] The objective of this report was to fill a critical gap: the absence of agreed-upon projections of future emissions. While the literature was stocked with hundreds of scenarios, there was no agreed-upon basis for evaluating the relative plausibility of future emissions scenarios. Nor were there any means to compare the scenarios in a way that would contribute to understanding the underlying assumptions of each in terms of socioeconomic factors such as population and economic growth, technological development, and lifestyle change.

By categorizing existing scenarios into classes on the basis of assumptions about the world and developing some new scenarios, *SRES* clarified a key contributing factor to the risk of climate change during the twenty-first century. While it is sometimes criticized for not quantifying the relative likelihoods of the scenarios it forwarded, *SRES* clearly advanced understanding of the relationships among scenarios, a critical step toward allowing individual experts to judge the probabilities (which occurred subsequently). *SRES* may not have amounted to basic research in social science, but it put the field in an epistemologically improved situation by increasing clarity about potential emissions pathways.

In sum, the IPCC embodies conflicting tendencies and inclinations. On the one hand is the official dictum against performing research. On the other is the perceived need for flexibility in developing information useful for policy makers. Sometimes this leads to inventiveness on the part of IPCC and its authors, of which *SRES* is an example. At other times, it leads to caution on the part of the authors, who in developing *AR4* leaned in the direction of avoiding performing research. The result was an ice sheet assessment that even some of the authors felt was not fully satisfactory and many observers felt was highly misleading.

RESEARCH AGENDAS, METHODOLOGICAL CHOICES, AND THE INTEGRATIVE FUNCTION OF ASSESSMENTS

Assessments strongly affect the production of knowledge by influencing research agendas and priorities. The controversial way IPCC assessors handled sea level projections, for example, spurred a raft of research and papers aimed at improving understanding of the problem. Some of these efforts were contemporaneous with *AR4* and were undertaken before the final result was in, and thus may not have been purely conceived as responses to the shortcomings of that assessment. Others were clearly and avowedly developed as a response.[9] Either way, in the following five years many papers appeared in professional journals elaborating new methods for estimating the ice sheet contribution in the absence of valid and comprehensive process-based models. This response shows one way in which the IPCC exerts leverage on the community's research agenda.

Another route of influence on the development of new knowledge is also demonstrated by the ice sheet example. It is common for authors of IPCC chapters to write papers during the assessment process on the subject they are assessing. These papers are reviewed and, if successful, published and then assessed, if not in time for the current IPCC assessment then for future assessments, sometimes by some of the same experts who wrote them. Inevitably, a strong back-and-forth develops between the gap-filling exigencies of the assessment and the research done by some of the authors in their work outside the immediate IPCC assessment context. Thus IPCC creates new knowledge by mobilizing the research community, not only prospectively but also contemporaneously.

One of the *AR4* authors of a chapter relevant to the ice sheet case, Stefan Rahmstorf, noting the inadequacy of existing modeling approaches, developed a new and controversial way to project sea level rise based on actual observations of recent and past sea level, called "semi-empirical modeling." During the *AR4* assessment, Rahmstorf suggested that his method be used as one source of evidence in developing the sea level projections, but he was rebuffed by the authors of the projection chapter. Participants stated that the method was rejected not because the idea arose directly from Rahmstorf's IPCC participation but because it did not involve a process-based model. Rahmstorf went ahead and published the method and the projections based on it in a series of now frequently-cited articles, the first of which was published almost simultaneously with *AR4*.[10] During development of the *Fifth Assessment Report* (*AR5*), the issue of the utility of semi-empirical modeling of sea level rise arose again and this time was discussed extensively in the chapter. However, even though Rahmstorf's approach was now available in the peer-reviewed literature, the chapter authors decided once again that the method did not provide a useful basis for projection due to the absence of process-based information. While still drafting their *AR5* chapter, several of the chapter authors published their reasoning in a peer-reviewed journal article.[11] The resulting IPCC chapter then drew on this article to make the argument against projecting sea level rise with semi-empirical models.

Assessments can also help to define research agendas by reformulating research questions and either answering them or stimulating research communities to do so. For example, climate models generate projections of daily maximum temperatures. Very high temperatures are called "extremes," and a heat wave is characterized as a cluster of days in which there are extreme temperatures (generally a period of three or more days when the maximum temperature exceeds some location-dependent value, e.g., 90°F or 95°F in the US northeast). Scientists studying the physics of climate are primarily interested in daily maximum temperatures, while climate impact experts and public health officials want to know about the likelihood of heat waves and their effect on human health. So the IPCC assesses both likelihoods of heat waves and likelihoods of extremely hot days. The same data from models and observations is assessed, but different questions are asked (and answered), useful to two distinct communities. By projecting the future chances of heat waves, the IPCC reframes existing knowledge about daily extremes developed by one expert community and produces new knowledge for another.

The influence of IPCC assessments on research questions goes even deeper. The assessments rely on a uniform set of projections of future climate in order to examine questions such as how the frequency and intensity of heat waves will change in the future, how many additional deaths might occur as a result, and what the cost of overall damage attributable to climate change later in this century will be. Developing these scenarios requires multiple runs of the more than three dozen available climate models.

The job of producing and evaluating these projections for the IPCC consumes much of the research time of scientists at key laboratories (generally governmental facilities) worldwide during each six-year IPCC cycle, a massive dedication of human and computer resources. This raises the issue not merely of the degree to which the IPCC determines allocation of people and resources but the degree to which questions not of interest to the IPCC are displaced from the global climate research agenda. For example, the focus on projections may come at the expense of improving understanding of the basic processes involved and implementing them in the model. In this respect, the IPCC plays an important role in producing new knowledge in some areas but also plays a role in inhibiting knowledge production in other areas.

The interests and exigencies of scientists in government laboratories also direct and constrain the questions the IPCC asks and the answers it offers.

For instance, with regard to scenarios of greenhouse gas emissions and concentrations in the atmosphere, climate scientists have in the past resisted requests to use a larger range of emissions scenarios, which limited the climate scenarios produced and the range of impacts that the IPCC was able to assess.

Stepping back from these particular ways of creating, extending, or limiting knowledge, we see that assessments also produce value by integrating existing knowledge, often by breaking disciplinary boundaries. The NRC assessment of the linearity question and the post–ozone hole redeployment of ozone depletion potentials created a new synthesis and new understandings from existing knowledge. The IPCC's Fourth Assessment started down a similar path in estimating the WAIS contribution to future sea level but never quite arrived at the destination.

In summary, assessments are not independent overviews comprising only the results of previously published work. They are the products of highly interactive processes that continually shape the questions asked and the answers given, where the groups of assessors and the groups of scientists whose work is assessed not only overlap strongly but actively work to redefine knowledge during the processes of assessment.

CONCLUSION

Scientific assessments do not just summarize existing knowledge; they also produce new knowledge. Institutional differences play an important role in how, when, and to what extent this occurs. NAPAP's development of "new knowledge" was guided and constrained by the particular exigencies of the agencies and their prerogatives within the NAPAP governing structure. When an outside assessor, the NRC, examined the linearity issue, it took advantage of its relative freedom to fill a knowledge gap, providing useful if incomplete insights that were timely in terms of the needs of the policy process. NAPAP used a different strategy to address the same issue, which relied on a multiyear large-scale modeling program that never attained its expected policy relevance. The IPCC took a broad view of the charge from its member governments to assess emission scenarios (*SRES*), but without specific guidance on ice sheets its authors made only a passing and unsatisfactory attempt to fill a critical knowledge gap. If IPCC member governments had asked for a special report on sea level rise, as they had asked for a special report on emissions scenarios, the outcome would likely have been quite different, and a more substantial body of new knowledge might have been created.

Governments strongly influence directions taken by assessments, but assessments also influence research agendas. This is especially clear in contemporary climate science: the IPCC exerts a strong influence on the research agendas of government laboratories. While pushing research in the direction of creating new knowledge for the policy world, it also constrains the issues examined to those that are of direct usefulness to IPCC assessments. This influence reaches well beyond scientists working for governments. There also is risk in such an arrangement: it entrains the possibility of scientists as a group running off in an unhelpful direction (i.e., "negative learning").[12]

To the extent that policy makers rely on an assessment, they are also relying on presumed self-correcting mechanisms within science. The hope is that scientists working in the field will recognize the shortcomings in an assessment report and deepen the body of relevant knowledge through new research. Events do unfold this way sometimes, as in the case of ice sheet assessment. However, there is no guarantee that this will occur. Moreover, if an assessment judges an issue to be unimportant—as in the assessment of heterogeneous reactions and ozone depletion—scientists may neglect something that is in fact important.

Many people think of the assessment process as a kind of summation—a critical or evaluative summary—of the existing knowledge in a particular field or group of fields pertinent to a particular policy problem. Many of the scientists we interviewed expressed this view, and some scientific assessors explicitly disavow the goal of creating new knowledge. On this view, assessors summarize, judge, and communicate policy-relevant knowledge to sponsors and stakeholders. From this perspective, the process can be viewed as a filter or screen that sends forward knowledge that is actionable, or, as Charles Kennel and Sally Daultrey have put it, "decision-ready."[13] The presumption is that the report is a critical summary of decision-ready knowledge; knowledge that does not meet that standard is screened out. There is no doubt that assessments attempt to do this, and most information in an assessment may fit the description of "decision-ready." But if some of the knowledge in the assessment is new, then this view of the assessment process is incorrect, or at least incomplete.

This reliability of new knowledge mobilized in an assessment may be further compromised if it is based on reports or other literature that have not been subject to peer review. Peer-reviewed literature may be of variable quality, but at least it has passed one established level of scrutiny, and the reputation of a journal generally bears some relation to the quality of the papers published in it, so assessors can judge both the paper itself and its provenance. The inclusion of other sorts of materials, often called "gray literature," puts an additional burden on the assessors to judge the quality of these materials. At minimum, this means there is a higher level of uncertainty associated with assessments that use materials that have not been subject to peer review than with those that restrict themselves to peer-reviewed materials only.

In the IPCC, there has been some confusion about this issue, as illustrated by the fracas caused by the error in the report of Working Group II in the *Fourth Assessment Report* on the melting of Himalayan glaciers.[14] While no report can ever be guaranteed to be error-free—and the significance of this particular error was exaggerated by some parties for political advantage—part of the story involved the use of a source that had not been peer reviewed.[15] The ensuing discussion brought to the fore the fact that IPCC scientists in some cases rely on reports of government agencies and nongovernmental organizations that have not been subject to formal scientific peer review. This occurs, particularly in Working Groups II and III, in part because some of the pertinent questions are not adequately addressed in the peer-reviewed literature, so scientists rely on what is available. But this means that to some extent they have mixed apples and oranges—or aged cabernet with Beaujolais nouveau—and the recipient of this knowledge may not have an entirely clear idea of these differences.

The issue of novelty also relates to the matter of consensus. If assessments were simply exercises in reviewing and filtering existing literatures, then consensus could be easily achieved by identifying and reporting overlapping knowledge claims in the peer-reviewed literature. But assessments also have a transformative and creative function. Assessors often add to the literature during the time they are assessing, and they influence the research trajectory of the domain that they are assessing. Assessment is part of the evolution of science and a driver of science, not simply a snapshot. The consensus forwarded in most scientific assessments is not a found object but also the product of human creativity.

Nor are scientific assessments simply a sink of scientific capital or a form of altruism through which scientists contribute to society but not to science. They are a form of scientific production. If the goal of science is to produce knowledge, then scientific assessments should be seen as one route toward achieving this goal, whether or not they succeed in satisfying the policy-related objectives for which they are typically established.

Conclusion

INTRODUCTION

We have examined the practice of scientific assessment, focusing on areas in the earth and environmental sciences in which landmark assessments have been carried out: acid precipitation, ozone depletion, and the fate of the West Antarctic Ice Sheet (WAIS). While modern scientific assessments differ from earlier attempts by experts to sort out troubling and contentious issues, assessment remains at its heart a process of discernment in which experts gather and judge evidence and attempt to discriminate among diverse, competing, and sometimes conflicting claims.[1] Assessments are convened for various purposes: sometimes they are meant to contribute to policy, sometimes to delay policy, sometimes to contribute to developing science in a particular area. Sometimes purposes are multiple, overlapping, or even antagonistic. Moreover, the functions that assessments actually fulfill may be different from what those who convened them intended. Here we summarize key conclusions from our work.

ASSESSMENTS ARE NOT JUST SUMMARIES

Assessments are often presented or discussed by scientists as if they were simply summaries or syntheses of existing knowledge and do not create new knowledge. For example, the emphasis in the Intergovernmental Panel on Climate Change (IPCC) on relying largely on peer-reviewed publications contributes to this impression.[2] We find that assessments are not just summaries but active sites of epistemic intervention. They construct as well as synthesize knowledge. This can occur when scientists grapple with uncertainties, conflicts, or contradictions in existing scientific information, or when they lack adequate information to characterize key elements. Assessments may also set the agenda for ongoing research, such as when the process of reviewing existing work identifies areas where available knowledge is insufficient to answer important policy-relevant questions. In these ways assessments can become sites of scientific knowledge production. This finding is important for at least three reasons.

First, the role of assessment in knowledge production is rarely recognized explicitly, and sometimes denied, as in the IPCC's "no research" rule. Rather than denying the knowledge-producing elements of assessments, thereby misrepresenting the character of the final product, it would be preferable to recognize the various forms of knowledge production embodied in assessments and to encourage those that enhance the value of the assessments to policy makers and perhaps discourage those that would be better undertaken in other settings (chapter 6).

Second, it is important to recognize that when scientists do produce new knowledge in an assessment, they find themselves in the position of assessing their own work. This is a challenge within assessments generally, since assessors are almost always contributors to the scientific literature they are assessing. Scientists are chosen to participate in assessments on the basis of their expertise, which is generally judged by the quantity and quality of their published papers in the area being assessed.

Third, assessments are advertised as reflecting established, agreed-upon knowledge, but new or evolving knowledge claims expressed in an assessment go beyond settled science. When the work being reviewed has been in the published literature for some time, there will be some independent sense of its significance and credibility based on how it has been received by the rest of the expert community, but when the work has emerged as part of the assessment itself, the assessors are in a weaker position to judge its credibility.

In addition to creating new knowledge, assessments can also act to create or sustain areas of ignorance.[3] One way they do this is by exclusion: topics may be knowingly excluded because scientists do not know how to handle them (e.g., dynamic ice sheet loss in the WAIS, chapter 4), or they may be ignored because scientists have (wrongly) concluded that they do not matter (e.g., heterogeneous reactions in stratospheric ozone depletion, chapter 3). Ignorance may also be created or sustained when experts who are perceived to have been too outspoken are marginalized or excluded, despite their having greater knowledge of the issue at hand than others who are included (chapter 5).

ASSESSORS DEFINE OBJECTIVITY AS POLICY NEUTRALITY AND A BALANCE OF BIAS

Assessments are convened by institutional actors in response to a perceived issue—a threat, conflict, or problem. The problem can be as universal and serious as the possibility of the elimination of all life on earth, as parochial as the desire to gain funding for a particular field of research, or as prosaic as helping political actors manage a controversial issue. Left to their own devices, many scientists would just do research in their laboratories or in the field without regard to social or political significance and without the added labor of communicating beyond their specialist communities. Assessments can therefore be understood as responding to a need created by the social organization of scientific life, offering knowledge that responds more directly to the needs of governance than is generally the case with conventional academic (or even some government-based) scientific research.[4]

Perhaps for this reason, many scientists see social engagement as a potential threat to scientific objectivity. Scientists often view the purpose of assessments—to serve the goals of governance and help to enact policy interventions—as in tension with the scientific norms of disinterestedness, dispassion, and social disengagement. In particular, they worry about maintaining objectivity in the face of these goals. Aware of the performative aspect of assessments, scientists also often believe that an assessment must be viewed as objective if it is to achieve the public purposes for which it is constituted.[5] For these reasons scientists involved in assessments often feel the need to articulate a notion of objectivity that is adequate to the particularities of the context.

Objectivity in the context of assessments is sometimes understood to operate on two levels, one individual and the other institutional. On the individual level, it is often construed as policy neutrality. As already noted, experts who have taken policy positions regarding the topic under consideration are often viewed as biased and may be excluded. On the institutional level, objectivity is regarded as a group or team achievement, and so there is an attempt to include individuals representing a variety of viewpoints, as well as a range of demographic characteristics such as race, gender, nationality (with the IPCC particularly concerned about including participants from both rich and poor nations), and type of employment (i.e., including experts from industry and nongovernmental advocacy organizations as well as from academia and government). Objectivity is viewed as a product of error cancellation: scientists' personal views, perspectives, and social and institutional locations and identities are viewed as potential sources of error that can be canceled by the presence of complementary sources of error. We call this approach a "balance of bias."

What needs to be balanced, on this view, are explicit policy commitments. Scientists are of course also citizens, who may hold private opinions that dispose them toward particular policies on matters related to their scientific work. However, these opinions are not generally regarded as threats to objectivity or even as relevant to scientists' participation in an assessment so long as they remain private. This creates an institutional disincentive for scientists to speak out on policy questions (at least so far as they might wish to participate in assessments).

Scientists may also hold strong methodological views—about the interpretation of data, about the best analytical approaches, or about other matters that are understood to be "internal" to science. These preferences may also be a source of bias, but they are treated inconsistently.[6] We have observed cases in which individuals with definitive scientific or methodological views were excluded (chapter 3), other cases where a range of views was sought (chapter 4), and still others in which the issue of bias was not addressed at all. Yet, as we saw in chapter 4, methodological preferences can affect outcomes as much as or more than policy preferences.

INSTITUTIONAL ARRANGEMENTS AFFECT EPISTEMIC OUTCOMES

Many natural scientists assume that institutional arrangements do not affect epistemic outcomes. We find this is not the case: institutional arrangements affect outcomes, sometimes dramatically. As already noted, assessments are commissioned by institutions: governments and their agencies or nongovernmental organizations. In every assessment, choices are made about who is invited to participate and who is not, how to define and frame the issues, what terms and conventions are used to characterize the problem, and how to carve up the questions into manageable pieces. These decisions are sometimes explicit, sometimes implicit, and not always made by scientists. Individuals working in assessment institutions may be scientists, but they may also be professional bureaucrats or political actors.

Assessments begin with a charge that may be relatively narrow (e.g., many assessments of the US National Research Council) or relatively broad (e.g., the US National Acid Precipitation Assessment Program [NAPAP] and the IPCC). The scope of an assessment affects questions asked and the resulting answers.

The charge helps to determine which experts are invited to participate in the assessment and which are not. Expertise is itself a matter of expert judgment; there are no formal rules for defining who counts as a relevant expert, and tacit assumptions about expertise may cause some experts to be overlooked or deliberately excluded. We have already mentioned how experts who have expressed policy views—or even the view that there should be a policy on the matter at hand—may be excluded as "biased." But other considerations lead to exclusions as well. Social scientists, for example, are often missing from assessments of issues that have important social dimensions. Even when social scientists are included, they often constitute a small fraction of the group.[7]

Moreover, insofar as assessments address complex, multidimensional problems, traditional markers of disciplinary expertise may be overvalued at the expense of other forms of expertise. As noted in the previous section, other factors—personal, institutional, geographical, epistemological—can also affect the constitution of assessment teams, and the manner in which these determinations are made can affect an assessment's conclusions (chapter 5).[8]

Assessments adopt particular conventions and vocabularies that affect how a problem is characterized. As we saw in chapter 2, the Canadian acid rain program set the pH reference value for acidification at 6, while the NAPAP Interim Assessment chose pH 5. Since one step down on the pH scale means a tenfold increase in acidity and significantly decreases the number of lakes recognized as acidified, many lakes that the Canadians would consider to be acidified disappear from concern in the NAPAP Interim Assessment. This definitional choice, which did not reflect an empirical disagreement, made the problem seem smaller than it might otherwise have seemed and thereby influenced both subsequent scientific work and policy discourse.

The way an assessment report is organized may also affect judgments that are expressed in the report. This is an important finding, because many scientists would expect that, with regard to matters of fact, the organization of a report would affect the order in which those facts are presented, and therefore perhaps the emphasis that different facts are given, but would not affect scientists' judgments about those facts. In chapter 4 we saw that this expectation was not met: the manner in which the assessment was organized (omitting a chapter specifically addressing sea level rise) affected the result produced. The idea that social and institutional factors affect intellectual outcomes is a familiar finding in the science studies literature yet has not been broadly assimilated by scientists and others who participate in and lead scientific assessments.[9]

THE PERMANENT ASSESSMENT ECONOMY

Prior to the twentieth century, assessments were largely ad hoc: experts gathered to study a matter, offered their opinion in a report or other form, and then disbanded. In the twentieth century this changed. The IPCC, for example, exists as an institution that moves almost immediately from the completion of one assessment to the beginning of another. It has officers with fixed terms elected by the constituent governments and a permanent staff. (This election process adds an important element which we have not studied here.) The US National Academy of Sciences similarly has boards, committees, and full-time staff whose work continues from one assessment to the next. Assessments under the auspices of various boards and committees proceed on a continuous basis. (Indeed, the US National Academy of Sciences has at any point in time a queue of studies waiting to commence.) In this sense, we see the emergence of what might be labeled a "permanent assessment economy."[10]

In the period we have studied, we observe a trend toward increasing institutionalization and internationalization of assessments. This appears to be motivated in part by the desire to achieve perceived objectivity via inclusivity.[11] With increased institutionalization come more formal rules, and ever more elaborate review procedures for producing the final document (chapter 5). This may be understood in part as an effort to demonstrate the fairness and legitimacy of the process in the face of actual or anticipated critiques.

The permanence of the assessment process contributes to the role that assessments play in creating scientific research agendas. Any assessment may articulate research needs and influence ongoing scientific work, but an assessment that is a prelude to another assessment is likely to have greater influence than one that closes up shop after writing its final report. This is a key element in how assessments have become scientific institutions in their own right and a locus of scientific knowledge production.

ASSESSORS STRIVE TO SEPARATE SCIENCE AND POLICY

Most of the assessments that we studied focused on science and refrained from policy recommendations. The most explicit statement of this is the expressed goal of the IPCC to be "policy-relevant but not policy-prescriptive."[12] This requires IPCC assessments to define a domain of science that is distinct from policy yet pertinent to it. Most of the other assessments we have studied seem to have followed a similar rule, albeit often implicitly.

One reason why many scientists involved in assessments are concerned to separate science from policy is because they view the assessment as a scientific project and therefore articulate their motivations primarily in scientific terms (to learn, to meet other scientists, to get up to speed in their subject area). They minimize or deny policy motivation or interest and downplay the salience of the political context. But this does not mean that the assessment itself is not motivated by political and social concerns, nor that the attempt to demarcate science from its political context succeeds (or ever could succeed). Nevertheless, many scientists speak as if they believe that such a demarcation is necessary, desirable, and achievable. Indeed, they often seem preoccupied by the distinction between science and policy.

In contrast, the UK vaccination commission of 1885 (chapter 1) was expected to, and did, recommend policy, and some early US ozone assessments (chapter 3) made policy recommendations in ways that later international ozone assessments have avoided and that the IPCC now expressly forbids. While we do not have sufficient examples to draw firm conclusions, it seems to be the case that scientists involved in large international assessments in recent decades have been engaged more explicitly in what sociologist Thomas Gieryn famously labeled "boundary work"—the articulation of clear boundaries between science and nonscience.[13]

The attempt to demarcate science from policy is linked to scientists' concern with objectivity—both actual and perceived—which they often interpret as requiring them to abstain from making policy recommendations. This raises the question of why scientists participating in internationalized assessments in recent decades have felt the need to refrain from policy in a way that their predecessors did not. Part of the explanation probably involves broad social trends toward increased professionalization and specialization and the increasingly explicit construction of norms governing various professions. Some evidence also suggests that as assessments become larger and more international, participants may feel a stronger need to demonstrate their independence and autonomy from (national or other) political pressures. One way to do this, it may be thought, is by articulating, respecting, and enforcing a clear separation between science and policy. It may also be the case that as the authority of science was increasingly challenged in the late twentieth century, scientists increasingly felt the need to demonstrate their objectivity in the hope that this would persuade others that they were worthy of trust.

No matter how strenuous the attempts by participants to disentangle science from policy, however, they are never entirely able to do so.[14] For one thing, the assessment itself is a policy intervention, and simply by participating in the assessments, scientists become enmeshed in a policy process. As pointed out in chapter 5, the policy response to a problem may be the assessment itself, which may have the effect of postponing or preventing other potential action. Moreover, even if scientists refrain from making formal policy recommendations in their report, that report is only one dimension of an assessment process that also includes interactions between experts and policy makers before, during, and after the publication of a report. As we saw in chapter 2, assessments may influence policy outcomes in other ways than through a written report.

The attempt to demarcate science and policy may in fact be undesirable, contrary to what many scientists believe, if it conflicts with or otherwise undermines the role of the scientist as sentinel. We have argued that scientists are sometimes in a privileged position to alert society to problems that would not otherwise be noticed, such as stratospheric ozone depletion (chapter 5).[15] If scientists are restrained from speaking in strong or even clear terms about problems of which only they have knowledge, this will compromise their ability to communicate potential threats, at great loss to society.

Our work supports the various science studies scholars who have rejected linear models of the relationship between science and policy, and we have noted the various ways in which they are entangled. Yet despite these insights, many scientists and others involved in creating and using assessments continue to believe that science and policy can and should be disentangled. We suggest that the demand to separate science from policy can be understood as a regulative ideal that can serve useful functions, including reminding scientists of the limits of their own expertise. We also suggest that scientists should not be reluctant to intervene in policy debates when their expertise is proximate but should be reluctant to intervene when it is not.[16]

ASSESSMENTS GENERALLY AIM FOR CONSENSUS

Many scientists believe that the goal of an assessment should be to offer a consensus statement. Consensus is not viewed simply as something that may (or may not) emerge from the process, but something toward which participants should actively work.

The drive for consensus likely has several motivations. One of these involves the performative role of assessments.[17] Assessments have audiences, and participants are aware of being watched. They also have a sense of being responsible, both to those who have commissioned the assessment and to fellow experts (although perhaps not so much to their fellow citizens, taxpayers, or some concept of society at large). This may lead scientists to want to speak univocally, feeling that expression of dissent will weaken the report. The thought is that univocality is requisite for influence. While the reward structure of academic life pulls toward disagreement, the demands of assessment push toward agreement (chapter 4).

The push toward agreement may be driven by a mental model that sees facts as what all reasonable people should be able to agree about versus differences of opinion or judgment that are potentially irresolvable.[18] If the claims in an assessment report are disputed, then it may be thought that they will be viewed as opinions rather than facts and thus dismissed not only by hostile critics but even by friendly forces. The drive toward consensus may therefore be an attempt to present the findings of the assessment as matters of fact rather than judgments of expert opinion.[19]

Our studies suggest that this impulse to achieve consensus can lead to undue conservatism—where "conservative" is understood to be reassuring rather than alarming conclusions—and "least common denominator" results reflecting the urge to find common ground. This may result in claims that are weak or incomplete (chapter 4). If consensus is viewed as a requirement, then scientists will put forward only those claims on which they all can fully agree. Thus scientists may avoid areas of controversy—even when these are recognized as scientifically significant—and underestimate potentially severe effects (what elsewhere we have labeled "erring on the side of least drama").[20] The desire for consensus may lead to not reporting certain views or to excluding certain experts. Similarly, the Global Environmental Assessment Project critiqued the consensus approach,[21] theorizing that authors of consensus assessments would tend to avoid issues that engender controversy among experts, and provided supporting evidence from a study about the way climate assessments dealt with WAIS.[22] In short, treating consensus as a goal may undermine other important goals, including inclusivity, accuracy, transparency, and clear communication.

We suggest that scientists should not view consensus as a goal of all assessments. Consensus should be viewed as an emergent property, not as something that necessarily needs to be achieved and certainly not something that should be enforced. Where substantive differences of opinion remain, they should be acknowledged and not downplayed and the reasons for them explained (to the extent that they can be explained). We also suggest that scientific communities should be open to experimenting with alternative models for making and expressing group judgments and to researching ways in which policy makers interpret the various assessment findings that result.

The law may be instructive in this regard. Before John Marshall became chief justice of the US Supreme Court in 1801, judges would give their opinions serially, as is still the norm in the United Kingdom. With Marshall came "the opinion of the Court"—a consensus statement of all the justices. It was not until the Dred Scott decision in 1857 that written dissents became common, and over the last century and a half they have had a profound influence on the development of the law.[23] Just as statements of dissent and the reasons for them are an important resource in the law, so they may be in science.

Assessments do not need to express consensus in order to be useful. They can add value by supplying a range of interpretations of the body of scientific knowledge, and in some cases this is clearly the better choice. For example, the IPCC's Fourth Assessment might have provided more useful guidance to policy makers if it had focused on the wide potential range of the ice sheet contribution to sea level rise rather than putting forward a partial projection of sea level rise that was widely derided and quickly out of date (chapter 4).

UNCERTAINTY

Uncertainty is a major focus of many assessments—its identification, articulation, quantification, and, where possible, reduction. One reason for this focus may be that all science is uncertain to some extent, but when policy decisions need to be made, these uncertainties come to the fore. Since uncertainty is often used as a reason not to act, those who believe action is needed are motivated to reduce those uncertainties.

In the assessments we studied, reduction of uncertainty was a central objective but was approached by the authors with a range of urgency. In assessing the linearity issue, NAPAP characterized the uncertainty but, in contrast to the National Research Council panel, deferred attempts to reduce it, awaiting the findings of the Regional Acid Deposition Model (chapter 2). In the wake of discovery of the ozone hole, assessors grappled with the difficulty of reducing the vast uncertainty in future ozone depletion absent a valid model describing the processes responsible (chapter 3). Eventually, perhaps goaded by intense concern on the part of policy makers (and themselves), they repurposed various metrics of ozone depletion to accomplish the task. Scientists assessing WAIS in the IPCC's Fourth Assessment attempted but failed to characterize uncertainty in the rate of dynamical ice loss and thus the ice sheet contribution to sea level rise.

For assessment authors, characterizing uncertainty is a two-step process. First, assessment authors need to establish the range of views in the scientific literature on the precision and accuracy of answers to particular questions—that is the review aspect of assessment. Second, they need to apply their own individual and group judgment about which if any of these views, or their combination, to favor—that is the critical judgment or discernment aspect of assessment. This act of discernment was carried out differently in each of our cases. If reducing uncertainty is viewed by sponsors as a priority objective of an assessment, then its authors need to consider their approaches to this goal at the beginning, so it can inform the process throughout and be explicit and transparent.

FINAL THOUGHTS

Assessments are a major part of the contemporary scientific and policy landscape and are likely to remain so for the foreseeable future. We offer the following thoughts to governments, experts, and scholars for their consideration and future research.

First, our studies support the ample prior work in science studies that challenges idealized models of the science/policy relationship based on a demarcation between the scientific and the social functions of the assessment.[24] Reality is patently more complex. Among other things, the authors of assessment reports are typically mindful of how their work may be used, neglected, or abused in political contexts, and they anticipate and respond to this in various ways. It is important for assessors to recognize, and for future research to continue to explore, the entanglements of science and policy in assessments, particularly how expectations of public reception or rejection influence what scientists do.

Second, we find that scientists in assessments typically approach the problem of objectivity via a "balance-of-bias" approach, working to create panels that are diverse with respect to race, gender, nationality, and, to some extent, employment type. In effect, they are implicitly accepting the view that objectivity is a social accomplishment. However, it remains unclear whether they do this primarily for political reasons (believing that the assessment needs to be perceived as inclusive in order to be credible and garner trust) or for epistemic ones (that diverse panels produce better scientific results).

Important questions remain as to how diversity is understood, how diverse assessment panels really are, and how well the various voices on panels are actually heard.[25]

Third, we find that many scientists believe that for an assessment to be impactful, it must express a consensus view; we find substantial reasons to question the wisdom of this view. We suggest that it would be worthwhile for scientific communities involved in assessments to explore and develop better means of handling and communicating disagreement.

Finally, we find that the entire assessment business—what we call the "assessment economy"—has come to influence research agendas profoundly (chapters 4 and 6).[26] This can be conscious and deliberate, as we saw in chapter 4, when early assessments of the threat of large-scale ice loss from the WAIS explicitly defined a research agenda for the community. However, it may also be inadvertent, as we see especially with regard to the IPCC. By establishing key questions and highlighting areas for future research, the IPCC now directly and indirectly influences the research of thousands of scientists, the questions asked and answered, and thus much of the knowledge that is available to be assessed. Inevitably, some questions and areas of research that would have otherwise been pursued are relinquished, thereby contributing to the production of ignorance in areas lacking immediate policy concern.[27] Governments, the organizations running assessments, and scientists themselves should be sensitive to this and find ways to ensure both that science and the assessment process continue to be innovative and that the assessment economy does not crowd out other forms of scientific knowledge production.

Interviews

Richard Alley, interviewed by Jessica O'Reilly, San Francisco, CA, December 14, 2009
Anonymous senior EPA scientist, interviewed by Milena Wazeck, by telephone, July 18, 2012
Tim Barnett, interviewed by Jessica O'Reilly, La Jolla, CA, November 23, 2009
Chris Bernabo, interviewed by Milena Wazeck, by telephone, June 10, 2011
Guy Brasseur, interviewed by Keynyn Brysse, Boulder, CO, June 25, 2009
John Church, interviewed by Jessica O'Reilly, by telephone, March 26, 2009
Tony Cox, interviewed by Keynyn Brysse, Cambridge, UK, March 16, 2009
Tony Cox, seminar/group interview by Keynyn Brysse, Cambridge, UK, March 16, 2009
Paul Crutzen, interviewed by Keynyn Brysse, Mainz, Germany, March 11, 2009
Robin Dennis, interviewed by Milena Wazeck, by telephone, July 5, 2012
Hermann Engelhardt, interviewed by Jessica O'Reilly, Pasadena, CA, April 23, 2009
David Fahey, interviewed by Keynyn Brysse, Boulder, CO, June 24, 2009
Joseph Farman, interviewed by Keynyn Brysse, Cambridge, UK, March 16, 2009
Helen Fricker, interviewed by Jessica O'Reilly, La Jolla, CA, May 26, 2009

Jonathan Gregory, interviewed by Jessica O'Reilly, Reading, UK, July 14, 2009
Jeremy Hales, interviewed by Milena Wazeck, by Skype, June 13, 2011
Neil Harris, interviewed by Keynyn Brysse, Cambridge, UK, March 16, 2009
Neil Harris, seminar/group interview by Keynyn Brysse, Cambridge, UK, March 16, 2009
Charles Herrick, interviewed by Milena Wazeck, by telephone, February 17, 2011
Bruce Hicks, interviewed by Milena Wazeck, Oak Ridge, TN, July 12, 2011
Richard Hindmarsh, interviewed by Jessica O'Reilly, Cambridge, UK, July 20, 2009
Ray Hosker, interviewed by Milena Wazeck, Oak Ridge, TN, July 12, 2011
Phillipe Huybrechts, interviewed by Jessica O'Reilly, Brussels, Belgium, July 3, 2009
James Mahoney, interviewed by Milena Wazeck, Leesburg, VA, July 15, 2011
John Malanchuk, interviewed by Milena Wazeck, Washington, DC, July 14, 2011
Mack McFarland, interviewed by Keynyn Brysse, Wilmington, DE, July 31, 2009
Michael McIntyre, seminar/group interview by Keynyn Brysse, Cambridge, UK, March 16, 2009
Johannes Oerlemans, interviewed by Jessica O'Reilly, Utrecht, the Netherlands, July 6, 2009
Michael Oppenheimer, interviewed by Jessica O'Reilly, Princeton, NJ, December 1, 2008
Jonathan Overpeck, interviewed by Jessica O'Reilly, Tucson, AZ, November 20, 2009
Tony Payne, interviewed by Jessica O'Reilly, Bristol, UK, July 10, 2009
John Pyle, interviewed by Keynyn Brysse, Cambridge, UK, March 17, 2009
Stefan Rahmstorf, interviewed by Jessica O'Reilly, by telephone, August 30, 2010
Eric Rignot, interviewed by Jessica O'Reilly, Irvine, CA, February 23, 2010
F. Sherwood Rowland, interviewed by Keynyn Brysse, Irvine, CA, March 5, 2009
Jonathan Shanklin, seminar/group interview by Keynyn Brysse, Cambridge UK, March 16, 2009

Keith Shine, interviewed by Keynyn Brysse, by telephone, March 25, 2009
Susan Solomon, interviewed by Jessica O'Reilly, Boulder, CO, October 7, 2010
Richard Somerville, interviewed by Jessica O'Reilly, La Jolla, CA, November 2009
O. Brian Toon, interviewed by Keynyn Brysse, Boulder, CO, June 25, 2009
Adrian Tuck, interviewed by Keynyn Brysse, Boulder, CO, June 24, 2009
Slawek Tulaczyk, interviewed by Jessica O'Reilly, Santa Cruz, CA, October 2009
Robert Turner, interviewed by Milena Wazeck, Oak Ridge, TN, July 11, 2011
David Vaughan, interviewed by Jessica O'Reilly, Cambridge, UK, July 17, 2009
Robert T. Watson, interviewed by Keynyn Brysse, Norwich, UK, March 13, 2009
Robert T. Watson, interviewed by Keynyn Brysse, by telephone, September 28, 2011
Don Wuebbles, interviewed by Keynyn Brysse, by telephone, February 3, 2010

Notes

PREFACE

1. See, e.g., Mitchell et al., *Global Environmental Assessments.*

CHAPTER ONE

1. Caciola, *Discerning Spirits,* 33.
2. Caciola, *Discerning Spirits,* chapter 3.
3. Caciola, *Discerning Spirits,* chapter 6.
4. Our account of this case relies on Alder, *Engineering the Revolution.*
5. Alder, *Engineering the Revolution,* 40.
6. On "men of science," see Lucier, "Pure and Applied Science."
7. A comment on language is in order we use the language of assessment in the sense defined here, but scientists participating in these activities sometimes refer to the written product as the assessment.
8. Farley and Geison, "Science, Politics and Spontaneous Generation."
9. Frankland, *Pasteur,* 62.
10. Hammonds, *Childhood's Deadly Scourge.*
11. Hammonds, *Childhood's Deadly Scourge,* chapter 4.
12. Royal Commission, *A Report on Vaccination and Its Results;* Wolfe and Sharp, "Anti-Vaccinationists Past and Present"; Krieger and Birn, "Vision of Social Justice as the Foundation of Public Health."
13. "Royal Commission on Vaccination,", 129.
14. Wolfe and Sharp, "Anti-Vaccinationists Past and Present"; "Report of the Royal Commission on Vaccination."
15. On the distinction, real or asserted, between science and policy, see Sarewitz, *Frontiers Of Illusion;* Pielke, *Honest Broker;* Parson, *Protecting the Ozone Layer;* and our discussion of the IPCC in chapter 4.
16. The "Dissentient Note," found on pp. 337–493 of the original report, is summarized in the article "Royal Commission on Vaccination." On Collins, see *Vaccination Inquirer and Health Review,* vol. 3.
17. "Report of the Royal Commission on Vaccination."
18. Winter, *Mesmerized.*
19. Harrington, *Medicine, Mind, and the Double Brain,* 28.

20. Winter, *Mesmerized*. On boundary work see Gieryn, "Boundary-Work."

21. Mesmerism and hypnosis were the subject of repeated assessments by the French Académie des Sciences of the work of Jean-Martin Charcot, whose work on animal magnetism had three times been the subject of condemnation by the académie. However, in 1882 he appeared before the académie again, this time describing his experiments with hypnotism. He outlined to the académie the stages of hypnotism and his attempts to use it as a treatment for hysteria. It is beyond the scope of this chapter to analyze why he succeeded in the end after three prior attempts; for a detailed discussion of this complex story, see Harrington, *Medicine, Mind, and the Double Brain*.

22. Porter, *Trust in Numbers*; Porter, *Karl Pearson*.

23. Mumford, *Technics and Civilization*.

24. This is a case where the expert evaluation explicitly included research and with it the creation of new knowledge; see chapter 5.

25. Burke, "Bursting Boilers and the Federal Power."

26. Crosland, *Science under Control*.

27. It also corresponds with what Will Steffen has labeled "The Great Acceleration"—the dramatic increase in population, pollution, and use of natural resources after World War II. The Anthropocene working group has identified this as a proposed beginning of the Anthropocene period; see Zalasiewicz et al., "When Did the Anthropocene Begin?"

28. Forman, "Behind Quantum Electronics"; Kevles, "Cold War and Hot Physics"; Krige, *American Hegemony*; Hamblin, *Arming Mother Nature*; Oreskes and Krige, *Science and Technology in the Global Cold War*; Oreskes, *Science on a Mission*.

29. Price, *Little Science, Big Science*; Weinberg, *Reflections on Big Science*; Galison and Hevly, *Big Science*.

30. Balogh, *Chain Reaction*, 22.

31. Doel, "Constituting the Postwar Earth Sciences"; Oreskes, *Science on a Mission*; Oreskes and Krige, *Science and Technology in the Global Cold War*.

32. Z. Wang, *In Sputnik's Shadow*; Mukerji, *Fragile Power*.

33. Dupree, "Founding of the National Academy of Sciences."

34. Frank Jewett, president of the Academy from 1939 to 1947, oversaw the initial transformation of the National Research Council. The emerging philosophy of the National Research Council is evident in his writings: e.g., Jewett, "The Academy."

35. It is interesting to note that Thomas Kuhn's seminal work, *The Structure of Scientific Revolutions*, which focused attention on the role of consensus in scientific communities, was published in 1962.

36. Kennel, "Global Knowledge Action Network."

37. "Univocality," as we are defining it here, refers to multiple experts essentially speaking with one voice. This is achieved through the common authorial practice of the expert assessment process, in which one report with one unified voice reports the consensus.

38. Krige, *American Hegemony*; Oreskes and Krige, *Science and Technology in the Global Cold War*.

39. In the United States, various commentators have noted the pressures in American society after World War II to seek unity in the face of profound racial, social, and economic differences, differences that had seemed to recede in the unifying atmosphere of the war but threatened to erupt again after it. Whyte, *Organization Man*; Mills, *Power Elite*. In the history of twentieth-century assessments, we see academic experts attempting, in their own way, to provide a basis for agreement about important social questions. It is a different matter, however, to establish that consensus statements are most useful to policy makers. It could well be that a Supreme Court style of decision making, which articulates the basis of dissent, is more useful in some contexts.

40. Even today there are various conceptions of what constitutes consensus and by what rules it should be achieved. Institutions and organizations that specialize in assessments—such as the Intergovernmental Panel on Climate Change and the US National Research Council—have developed their own formalized processes for consensus formation and consensus reporting.

41. Parson, *Protecting the Ozone Layer*.

42. Oreskes and Conway, *Merchants of Doubt*.

43. Brysse et al., "Climate Change Prediction."

44. This is similar to what Paul Edwards argues in his discussion of the mutual alignment of scientists and military patrons in computer science. Edwards, *The Closed World*.

45. Galison, *Image and Logic*.

46. Whether or not this is achieved is a different matter.

47. Roberts, "Learning from an Acid Rain Program"; Roberts, "Acid Rain Program"; Rubin, Lave, and Morgan, "Keeping Climate Research Relevant"; Oversight Review Board of the National Acid Precipitation Assessment Program, *Experience and Legacy of NAPAP*; Herrick, "Predictive Modeling of Acid Rain"; Herrick and Jamieson, "Social Construction of Acid Rain."

48. Lewandowsky, Risby, and Oreskes, "'Pause' in Global Warming."

49. Smithson, "Conflict Aversion."

50. Cabantous, "Ambiguity Aversion in the Field of Insurance."

51. US National Research Council, "About Our Expert Consensus Reports," http://dels.nas.edu /global/Consensus-Report (accessed September 11, 2016).

CHAPTER TWO

1. Mahoney, Malanchuck, and Irving, "Plan and Schedule for NAPAP's 1989 and 1990 Assessment Reports," 1492.

2. Roberts, "Learning from an Acid Rain Program"; Roberts, "Acid Rain Program"; Rubin, Lave, and Morgan, "Keeping Climate Research Relevant"; Oversight Review Board of the National Acid Precipitation Assessment Program, *Experience and Legacy of NAPAP*; Herrick, "Predictive Modeling of Acid Rain."

3. Roberts, "Federal Report on Acid Rain Draws Criticism."

4. Quoted in Shabecoff, "Government Acid Rain Report Comes under Sharp Attack."

5. This is especially true for Kulp, who came out with clear statements against acid rain legislation after he resigned as NAPAP's director. See Kulp, "Government Measures to Reduce Acid Rain Are Unnecessary."

6. R. Smith, "On the Air and Rain in Manchester," 213.

7. R. Smith, "On the Air and Rain in Manchester," 211.

8. R. Smith, "On the Air and Rain in Manchester," 213.

9. Likens and Bormann, "Acid Rain."

10. The NADP's website states that "compounds from industry in China can potentially be deposited in the US Midwest. For this reason, acid rain is considered a global problem." http://nadp .sws.uiuc.edu/educ/acidrain.aspx (accessed July 3, 2016).

11. Hales, *Tall Stacks and the Atmospheric Environment*; Likens, "Acid Rain."

12. Forster, *Acid Rain Debate*, 21.

13. Cowling, "Effects of Acid Precipitation and Atmospheric Deposition on Terrestrial Vegetation."

14. Cowling, "Effects of Acid Precipitation and Atmospheric Deposition on Terrestrial Vegetation," 56.

15. OECD, *OECD Programme on Long Range Transport of Air Pollutants*, 3.

16. Oden, "Nederbördens Försurning." A similar phrase was also used in the United States in an article written by the ecologists Gene Likens, Herbert Bormann, and Noye Johnson in 1972 ("Air pollution is an unpremeditated form of chemical warfare!"). Likens, Bormann, and Johnson, "Acid Rain," 40.

17. Wetstone and Rosencranz, *Acid Rain in Europe and North America*, 133.

18. Participating member countries were Austria, Belgium, Denmark, Germany, Finland, France, the Netherlands, Norway, Sweden, Switzerland, and the United Kingdom. OECD, *OECD Programme on Long Range Transport of Air Pollutants*; see also Rothschild, "Burning Rain."

19. OECD, *OECD Programme on Long Range Transport of Air Pollutants*, 1.

20. OECD, *OECD Programme on Long Range Transport of Air Pollutants*, 1.

21. Likens, Bormann, and Johnson, "Acid Rain"; Likens and Bormann, "Acid Rain."

22. US NAS, Panel on Atmospheric Chemistry, *Atmospheric Chemistry*; Dochinger and Selinga, *Proceedings of the First International Symposium on Acid Precipitation and the Forest Ecosystem*; Committee on Sulfur Oxides, *Sulfur Oxides*.

23. US Department of Energy, *Multistate Atmospheric Power Production Pollution Study—MAP3S*.

24. Overrein, Seip, and Tollan, *Acid Precipitation—Effects on Forests and Fish*, 3.

25. Acid Precipitation Act of 1980, Public Law 96-294, *US Statutes at Large* 94 (1980), 771.

26. Canada/United States Coordinating Committee. *United States-Canada Memorandum of Intent on Transboundary Air Pollution*, 1:xxxii.

27. The comparison draws on work by Schmandt, Clarkson, and Roderick, who compared 15 US and Canadian acid rain reports from the early 1980s. Schmandt, Clarkson, and Roderick, *Acid Rain and Friendly Neighbors*. The European assessments are included in order to address questions regarding the potential influence of nationality or locality on the evaluation of the causes and effects of acid deposition. There are, however, several limitations to this comparison: it does not include all acid rain assessments that were published during this period, and it focuses on selected aspects of acid rain research.

28. Overrein, "Report from the International Conference on the Effects of Acid Precipitation"; OECD, *OECD Programme on Long Range Transport of Air Pollutants*; Overrein, Seip, and Tollan, *Acid Precipitation—Effects on Forests and Fish*.

29. Galloway et al., *National Program for Assessing the Problem of Atmospheric Deposition*; NRC, *Atmosphere-Biosphere Interactions*; National Research Council of Canada, Associate Committee on Scientific Criteria for Environmental Quality, *Acidification in the Canadian Aquatic Environment*.

30. Interagency Task Force on Acid Precipitation, *National Acid Precipitation Assessment Plan Draft*.

31. Overrein, "Report from the International Conference on the Effects of Acid Precipitation"; NRC, *Atmosphere-Biosphere Interactions*.

32. Wetstone and Rosencranz, *Acid Rain in Europe and North America*, 95–96.

33. Clean Air Act Amendments of 1977, Public Law 95-95, *US Statutes at Large* 91 (1977), 693.

34. Clean Air Act Amendments of 1977, 710.

35. Canadian Clean Air Act, Section 21.2.

36. Douglas Costle to George Mitchell, January 13, 1981, reprinted in *New York v. Thomas*, app. A–B, 613 F. Supp. 1472, 1486, 1488 (1985).

37. One of the authors of this volume, Michael Oppenheimer, was a senior scientist at the EDF from 1981 to 2002 and participated in these debates.

38. Senator Moynihan, speaking on S. 1754, on September 14, 1979, 96th Cong., 1st Sess., *Congressional Record* 125, pt. 19, 24631–24632.

39. Cowling, in "Performance and Legacy of NAPAP," claims that NAPAP's origins reach even further back. In 1977, the CEQ asked four university scientists and leading experts in acid rain research (Ellis B. Cowling, James Galloway, Eville Gorham, and William McFee) to "assess the present scope of available knowledge about atmospheric deposition and its effect on agricultural lands, forests, ranges, parks, surface waters, and aquatic life in the United States" and "to recommend a coordinated program of research and monitoring necessary to serve as a basis for the intelligent management of atmospheric emissions and for the amelioration of the adverse effects of atmospheric deposition on plant and animal life." Galloway et al., *National Program for Assessing the Problem of Atmospheric Deposition*, 2.

40. Likens and Bormann, "Acid Rain," 1176.

41. Galloway et al., *National Program for Assessing the Problem of Atmospheric Deposition*, 36. Galloway's view changed during the next few years. In a congressional hearing in 1988, he urged

policy makers to take action: "The scientific community has accumulated enough information for the policymakers to stop the discussion about whether to reduce emissions of sulfur and nitrogen, and to begin the discussion of how to reduce them." Galloway, Testimony of James N. Galloway, 123. It is unclear how representative Galloway's change of views is. During the 1980s, scientists accumulated more knowledge about acid rain, but there was also disappointment regarding the Reagan administration's reluctance to address the problem.

42. Cowling and Nilsson, "Comparison of Some National Assessments."

43. We thank Robin Dennis for an interesting and helpful conversation about this point.

44. Chris Bernabo, telephone interview by Milena Wazeck, June 10, 2011.

45. NAPAP, *NAPAP Interim Assessment,* 1:2.

46. In fact, the task force consisted of the original Acid Rain Coordinating Committee members plus additional members. NAPAP, *National Acid Precipitation Plan,* 4. This interagency model was apparently modeled after the 1978 National Climate Program Act. Chris Bernabo, telephone interview by Milena Wazeck,, June 10, 2011, 6:10.

47. Interagency Task Force on Acid Precipitation, *National Acid Precipitation Assessment Plan Draft,* v.

48. NAPAP, *Annual Report to the President and Congress (1982),* 6.

49. Bruce Hicks, interview by Milena Wazeck, Oak Ridge, TN, July 12, 2011.

50. James Mahoney, interview by Milena Wazeck, Leesburg, VA, July 15, 2011.

51. John Malanchuk, interview by Milena Wazeck, Washington, DC, July 14, 2011.

52. John Malanchuk, interview by Milena Wazeck, Washington, DC, July 14, 2011.

53. NAPAP, *NAPAP Interim Assessment,* 1:2.

54. It is not possible to give a precise number of scientists involved. NAPAP documents speak of "approximately 300 authors" for the State of Science/Technology (SOS/T) reports (NAPAP, *Acid Deposition,* 1:iii), and even more scientists had their research funded through NAPAP but were not authors of the SOS/T reports.

55. NAPAP, *Annual Report to the President and Congress (1983),* 61, 63.

56. NAPAP, *Annual Report to the President and Congress (1983),* 61.

57. NAPAP, *Annual Report to the President and Congress (1983),* 62.

58. The development of the integrated assessment methodology is described as a major goal. Studies were in particular conducted by the National Laboratory Consortium and by Carnegie-Mellon University. NAPAP, *Annual Report to the President and Congress (1983),* 62.

59. Interagency Task Force on Acid Precipitation, *Report on the Second Annual Review Meeting of the National Acid Precipitation Assessment Program,* 1:381–419.

60. John Malanchuk, interview by Milena Wazeck, Washington, DC, July 14, 2011.

61. Likens, "Role of Science in Decision Making."

62. Domestic Policy Council (DPC), memorandum for the president, January 13, 1988, Howard Baker Files, box 1, Ronald Reagan Library.

63. Jack Svahn, memorandum for Edwin Meese III, September 12, 1983, Edwin Meese Files, box 70, folder OA 11863, Ronald Reagan Library.

64. Danny J. Boggs, "Cabinet Council on Natural Resources and Environment Working Group on Acid Rain Policy," July 21, 1983, Danny Boggs Files, box 4, folder OA 11461, Ronald Reagan Library.

65. DPC, memorandum for the president, January 13, 1988, Howard Baker Files, box 1, Ronald Reagan Library.

66. Al Hill et al., memorandum to Ed Meese et al., November 4, 1982, Dunlop Files, box 6, Ronald Reagan Library.

67. Edwin Meese Files, box 27, folder OA 9447, Ronald Reagan Library.

68. Council on Environmental Quality, memorandum for Working Group on Energy, Natural Resources and Environment, November 1, 1985, John Svahn Files, folder OA 13531, Ronald Reagan Library.

69. James Mahoney, interview by Milena Wazeck, Leesburg, VA, July 15, 2011.

70. NAPAP, *Annual Report to the President and Congress (1986)*, 19.

71. NAPAP, *Annual Report to the President and Congress (1983)*, 61.

72. J. Laurence Kulp, quoted in Roberts, "Federal Report on Acid Rain Draws Criticism," 1404.

73. US General Accounting Office, *Acid Rain*, 20.

74. US General Accounting Office, *Acid Rain*, 49.

75. NAPAP, *NAPAP Interim Assessment*, 1:13–16.

76. NAPAP, *NAPAP Interim Assessment*, 1:19.

77. Kulp, "Acid Rain"; Kulp, "Government Measures to Reduce Acid Rain Are Unnecessary."

78. US General Accounting Office, *Acid Rain*, 49.

79. Anonymous administration official, quoted in Roberts, "Federal Report on Acid Rain Draws Criticism," 1404.

80. Bruce Hicks, interview by Milena Wazeck, Oak Ridge, TN, July 12, 2011.

81. Ad Hoc Committee on Acid Rain: Science and Policy, *Is There Scientific Consensus on Acid Rain?*

82. US Congress (House), Subcommittee on Natural Resources, Agriculture Research, and Environment, Committee on Science, Space, and Technology, *National Acid Precipitation Assessment Program*, 3.

83. Tom McMillan, quoted in Peterson, "Downpour of Criticism Greets Acid-Rain Report.".

84. RMCC, *Critique of the US National Acid Precipitation Assessment Program's Interim Assessment Report*, 1.

85. Likens, Letter from Gene E. Likens to Congressman John D. Dingell, 65–71.

86. National Research Defense Council, quoted in Peterson, "Downpour of Criticism Greets Acid-Rain Report.".

87. NAPAP, *NAPAP Interim Assessment*, 1:I-8.

88. We were not able to locate this testimony. Bledsoe stated in his memorandum for Nancy Risk that Kulp would testify before Congress in the second week of July 1987. There was a hearing on the Acid Deposition Control Act of 1987 on July 9–10, 1987, but the congressional record did not include testimony by Kulp.

89. Bledsoe, memorandum for Nancy Risque, July 1, 1987, Nancy Risque Files, box 1, folder OA16066, Ronald Reagan Library.

90. Minutes of the DPC, August 5, 1987, Nancy Risque Files, box 1, folder OA 16066, Ronald Reagan Library.

91. Minutes of the DPC, August 5, 1987, Nancy Risque Files, box 1, folder OA 16066, Ronald Reagan Library.

92. Minutes of the DPC, September 22, 1987, Sprinkel Files, box 10, folder OA 17755, Ronald Reagan Library.

93. Bledsoe, memorandum for the DPC, September 21, 1987, Sprinkel files, box 10, folder OA 17755.

94. McMillan, Letter of Thomas McMillan to Lee Thomas, January 6.

95. James Mahoney, interview by Milena Wazeck, Leesburg, VA, July 15, 2011.

96. Canada/United States Coordinating Committee. *United States–Canada Memorandum of Intent on Transboundary Air Pollution*, vol. 2; US EPA, *Acid Deposition Phenomenon and Its Effects*, vol. 1; NRC, Committee on Atmospheric Transport and Chemical Transformation in Acid Precipitation, *Acid Deposition*; RMCC, *Assessment of the State of Knowledge.*

97. US EPA, *Acid Deposition Phenomenon and Its Effects*, vol. 1; NRC, Committee on Atmospheric Transport and Chemical Transformation in Acid Precipitation, *Acid Deposition*; OOTA, *Acid Rain and Transported Air Pollutants*; NAPAP, *NAPAP Interim Assessment*; RMCC, *Assessment of the State of Knowledge.*

98. NAPAP and RMCC, *Joint Report to the Bilateral Advisory and Consultative Group.*

99. NAPAP, *NAPAP Interim Assessment,* 1:I-10.

100. McMillan, Letter of Thomas McMillan to Lee Thomas, January 6, 140.

101. E.g., Harvey and Beamish, "Acidification of the La Cloche Mountain Lakes, Ontario"; Schofield, "Acid Precipitation."

102. NAPAP, *NAPAP Interim Assessment,* 1:I-8.

103. NAPAP, *1990 Integrated Assessment Report,* 11.

104. Schindler, "View of NAPAP From North of the Border," 125; Eville Gorham and Michael Oppenheimer, cited in Shabecoff, "Government Acid Rain Report Comes Under Sharp Attack"; RMCC, *Critique of the US National Acid Precipitation Assessment Program's Interim Assessment Report,* 1, 6; Likens, Testimony to the Subcommittee on Natural Resources, Agriculture Research, 57; US Congress (House), Subcommittee on Natural Resources, Agriculture Research, and Environment, Committee on Science, Space, and Technology. *National Acid Precipitation Assessment Program,* 45; EDF, Testimony of Joseph Goffman, 131–132.

105. RMCC, *Critique of the US National Acid Precipitation Assessment Program's Interim Assessment Report,* 5.

106. NAPAP, *NAPAP Analysis of the Critiques of NAPAP's 1987 Interim Assessment,* 13.

107. NAPAP, *NAPAP Analysis of the Critiques of NAPAP's 1987 Interim Assessment,* 13.

108. Likens, Testimony to the Subcommittee on Natural Resources, Agriculture Research, 57.

109. NAPAP, *NAPAP Interim Assessment,* 1:I-8.

110. McMillan, Letter of Thomas McMillan to Lee Thomas, January 6, 142; RMCC, *Critique of the US National Acid Precipitation Assessment Program's Interim Assessment Report,* 5; Likens, Testimony to the Subcommittee on Natural Resources, Agriculture Research, 57.

111. RMCC, *Critique of the US National Acid Precipitation Assessment Program's Interim Assessment Report,* 5.

112. Schindler, "View of NAPAP from North of the Border," 125.

113. OTA, *Acid Rain and Transported Air Pollutants,* 9.

114. McLaughlin, "Effects of Air Pollution on Forests."

115. NAPAP and RMCC, *Joint Report to the Bilateral Advisory and Consultative Group,* 7.

116. US EPA, *Acid Deposition Phenomenon and Its Effects*; NAPAP, *NAPAP Interim Assessment.*

117. NAPAP and RMCC, *Joint Report to the Bilateral Advisory and Consultative Group,* executive summary, 3.

118. NAPAP, *NAPAP Interim Assessment,* 1:I-9.

119. NAPAP, *NAPAP Interim Assessment,* 1:I-28, 4:7–51.

120. Likens, Testimony to the Subcommittee on Natural Resources, Agriculture Research, 58; EDF, Testimony of Joseph Goffman, 132; RMCC, *Critique of the US National Acid Precipitation Assessment Program's Interim Assessment Report,* 2, 21.

121. Paoletti et al., "Advances of Air Pollution Science."

122. NAPAP, *NAPAP Interim Assessment,* 1:I-8.

123. The other three were the neglect of the relative importance of meteorology versus chemistry, low-elevation hardwood forest dieback, and Canadian modeling and survey results on aquatic effects. See RMCC, *Critique of the US National Acid Precipitation Assessment Program's Interim Assessment Report,* 4.

124. RMCC, *Critique of the US National Acid Precipitation Assessment Program's Interim Assessment Report,* 3; see also Galloway, Testimony of James N. Galloway, 120.

125. RMCC, *Critique of the US National Acid Precipitation Assessment Program's Interim Assessment Report,* 4.

126. RMCC, *Critique of the US National Acid Precipitation Assessment Program's Interim Assessment Report,* 14.

127. NAPAP, *NAPAP Analysis of the Canadian RMCC Critique of the NAPAP Interim Assessment,* 6.

128. Quoted in US Congress (House), Subcommittee on Natural Resources, Agriculture Research, and Environment, Committee on Science, Space, and Technology, *National Acid Precipitation Assessment Program*, 43.

129. NAPAP, *NAPAP Analysis of the Canadian RMCC Critique of the NAPAP Interim Assessment*; NAPAP, NAPAP *Analysis of the Critiques of NAPAP's 1987 Interim Assessment.*

130. Quoted in US Congress (House), Subcommittee on Natural Resources, Agriculture Research, and Environment, Committee on Science, Space, and Technology, *National Acid Precipitation Assessment Program*, 38.

131. NAPAP, *NAPAP Interim Assessment*, 1:I-8.

132. NAPAP, *NAPAP Interim Assessment*, 1:7-1.

133. NAPAP, *NAPAP Interim Assessment*, 1:I-8.

134. NAPAP, *NAPAP Interim Assessment*, 3:4-2.

135. NAPAP, *NAPAP Interim Assessment*, 3:4-32.

136. NAPAP, *NAPAP Interim Assessment*, 1:I-9.

137. NAPAP, *NAPAP Interim Assessment*, 4:6-44.

138. NAPAP, *NAPAP Interim Assessment*, 1:I-9.

139. NAPAP, *NAPAP Interim Assessment*, 4:7-1.

140. NAPAP, *NAPAP Interim Assessment*, 4:7-27.

141. McMillan, Letter of Thomas McMillan to Lee Thomas, January 6, 140.

142. RMCC, *Critique of the US National Acid Precipitation Assessment Program's Interim Assessment Report*, 1.

143. RMCC, *Critique of the US National Acid Precipitation Assessment Program's Interim Assessment Report*, 1–3.

144. Galloway et al., *National Program for Assessing the Problem of Atmospheric Deposition*, 1.

145. NRC, *Atmosphere-Biosphere Interactions*, 3.

146. Brysse et al., "Climate Change Prediction."

147. NAPAP, *NAPAP Analysis of the Canadian RMCC Critique of the NAPAP Interim Assessment*, 1.

148. Acid Precipitation Act of 1980, Public Law 96-294, *US Statutes at Large* 94 (1980), 771.

149. Interagency Task Force on Acid Precipitation, *National Acid Precipitation Assessment Plan Draft*, v.

150. Interagency Task Force on Acid Precipitation, *Report on the Second Annual Review Meeting of the National Acid Precipitation Assessment Program*, 1–5.

151. NAPAP, *Annual Report to the President and Congress (1983)*, 1.

152. NAPAP, *Annual Report to the President and Congress (1983)*, 6.

153. Jeremy Hales, Skype interview by Milena Wazeck, June 13, 2011.

154. John Malanchuk, interview by Milena Wazeck, Washington, DC, July 14, 2011.

155. James Mahoney, interview by Milena Wazeck, Leesburg, VA, July 15, 2011.

156. James Mahoney, interview by Milena Wazeck, Leesburg, VA, July 15, 2011.

157. James Mahoney, interview by Milena Wazeck, Leesburg, VA, July 15, 2011.

158. NAPAP, *Plan and Schedule for NAPAP Assessment Reports, 1989–1990*, vii, emphasis in original.

159. NAPAP, *1990 Integrated Assessment Report*, 2–4.

160. NAPAP, *1990 Integrated Assessment Report*, i.

161. NAPAP, *1990 Integrated Assessment Report*, 515.

162. NAPAP, *1990 Integrated Assessment Report*, 519.

163. John Malanchuk, interview by Milena Wazeck, Washington, DC, July 14, 2011; see also C. Mann, "Meta-Analysis in the Breech."

164. John Malanchuk, interview by Milena Wazeck, Washington, DC, July 14, 2011.

165. John Malanchuk, interview by Milena Wazeck, Washington, DC, July 14, 2011.

166. NAPAP, *1990 Integrated Assessment Report*, 407.

167. C. Mann, "Meta-Analysis in the Breech," 478.

168. Roberts, "Going for Broke on a Mega Model," 1304.

169. NAPAP, *Models Planned for Use in the NAPAP Integrated Assessment*, 4-10.

170. NAPAP, *Models Planned for Use in the NAPAP Integrated Assessment*, 1-1.

171. Forster, *Acid Rain Debate*, 25; White, *Acid Rain*; US EPA, *Acid Deposition Phenomenon and Its Effects*, 1:4-72.

172. Quoted in US Congress (Senate), Committee on Environment and Public Works, *Review of the Federal Government's Research Program on the Causes and Effects of Acid Rain*, 14–15.

173. Brimblecombe, *Big Smoke*; Uekoetter, *Age of Smoke*.

174. Hales, *Tall Stacks and the Atmospheric Environment*.

175. Pratt and Pratt, "Acid Rain."

176. NAPAP, *Acid Deposition*, 1:2-248.

177. OECD, *OECD Programme on Long Range Transport of Air Pollutants*; Overrein, Seip, and Tollan, *Acid Precipitation—Effects on Forests and Fish*.

178. Bruce Hicks, interview by Milena Wazeck, Oak Ridge, TN, July 12, 2011.

179. NRC, Committee on Atmospheric Transport and Chemical Transformation in Acid Precipitation, *Acid Deposition*, 139.

180. NAPAP, *Annual Report to the President and Congress (1983)*, 14.

181. Robin Dennis, telephone interview by Milena Wazeck, July 5, 2012.

182. NAPAP, *NAPAP Interim Assessment*, 3:4–65, 3:4–67.

183. NAPAP, *1990 Integrated Assessment Report*, 172.

184. NAPAP, *Acid Deposition*, 1:2-41.

185. John Malanchuk, interview by Milena Wazeck, Washington, DC, July 14, 2011.

186. Bruce Hicks, interview by Milena Wazeck, Oak Ridge, TN, July 12, 2011.

187. Jeremy Hales, Skype interview by Milena Wazeck, June 13, 2011.

188. NAPAP, *Models Planned for Use in the NAPAP Integrated Assessment*, 4-2.

189. NAPAP, *Models Planned for Use in the NAPAP Integrated Assessment*, 4-2.

190. US General Accounting Office, *Air Pollution*, 23.

191. NAPAP, *Models Planned for Use in the NAPAP Integrated Assessment*, 4-13.

192. John Malanchuk, interview by Milena Wazeck, Washington, DC, July 14, 2011.

193. Jeremy Hales, Skype interview by Milena Wazeck, June 13, 2011.

194. John Malanchuk, interview by Milena Wazeck, Washington, DC, July 14, 2011.

195. Perhac, "Making Credible Science Usable," 38.

196. John Malanchuk, interview by Milena Wazeck, Washington, DC, July 14, 2011.

197. US General Accounting Office, *Air Pollution*, 4.

198. Robin Dennis, telephone interview by Milena Wazeck, July 5, 2012.

199. Rubin, "Benefit-Cost Implications of Acid Rain Controls," referring to NAPAP, *1990 Integrated Assessment Report*, questions 4 and 5, external review draft.

200. National Energy Technology Laboratory, IGCC Project Examples, https://www.netl.doe.gov/research/coal/energy-systems/gasification/gasifipedia/project-examples (accessed February 2, 2017).

201. "Cap and Trade: Acid Rain Program Results," https://grist.files.wordpress.com/2009/06/ctresults.pdf (accessed February 25, 2014).

202. NAPAP, *1990 Integrated Assessment Report*, 510.

203. NAPAP, *Plan and Schedule for NAPAP Assessment Reports, 1989–1990*, 18.

204. Bruce Hicks, interview by Milena Wazeck, Oak Ridge, TN, July 12, 2011.

205. John Malanchuk, interview by Milena Wazeck, Washington, DC, July 14, 2011.

206. Arrow Papers, box 13, Duke University, emphasis in original.

207. Russell, memorandum to ORB, February 23, 1989, Arrow Papers, Duke University, emphasis in original.

208. John Malanchuk, interview by Milena Wazeck, Washington, DC, July 14, 2011.

209. James Mahoney, interview by Milena Wazeck, Leesburg, VA, July 15, 2011.

210. John Malanchuk, interview by Milena Wazeck, Washington, DC, July 14, 2011; Bruce Hicks, interview by Milena Wazeck, Oak Ridge, TN, July 12, 2011.

211. Schindler, "View of NAPAP from North of the Border," 127; Cowling, "Performance and Legacy of NAPAP."

212. Schindler, "View of NAPAP from North of the Border," 128.

213. See, e.g., Likens, "Role of Science in Decision Making."

214. Roberts, "Learning from an Acid Rain Program"; Roberts, "Acid Rain Program"; Rubin, Lave, and Morgan, "Keeping Climate Research Relevant"; Oversight Review Board of the National Acid Precipitation Assessment Program, *Experience and Legacy of NAPAP*; Herrick, "Predictive Modeling of Acid Rain."

215. Rubin, "Benefit-Cost Implications of Acid Rain Controls"; Perhac, "Making Credible Science Usable"; Oversight Review Board of the National Acid Precipitation Assessment Program, *Experience and Legacy of NAPAP*.

216. James Mahoney, interview by Milena Wazeck, Leesburg, VA, July 15, 2011.

217. John Malanchuk, interview by Milena Wazeck, Washington, DC, July 14, 2011.

218. K. Schneider, "Lawmakers Agree on Rules to Reduce Acid Rain Damage."

219. James Mahoney, interview by Milena Wazeck, Leesburg, VA, July 15, 2011.

220. John Malanchuk, interview by Milena Wazeck, Washington, DC, July 14, 2011.

221. Messer, "Monitoring, Assessment, and Environmental Policy," 502.

222. David Hawkins, quoted in Roberts, "Going for Broke on a Mega Model," 1304.

223. James Mahoney, quoted in Roberts, "Going for Broke on a Mega Model," 1304.

224. Herrick and Jamieson, "Social Construction of Acid Rain," 110.

225. John Malanchuk, interview by Milena Wazeck, Washington, DC, July 14, 2011.

226. James Mahoney, interview by Milena Wazeck, Leesburg, VA, July 15, 2011.

227. With the exception of Ruckelshaus, and there were congressmen interested in what NAPAP was doing.

CHAPTER THREE

1. Litfin, *Ozone Discourses*; LePrestre, Reid, and Morehouse, *Protecting the Ozone Layer*; Christie, *The Ozone Layer*; Grundmann, *Transnational Environmental Policy*; Andersen, Sarma, and Sinclair, *Protecting the Ozone Layer*. See Betsill and Pielke, "Blurring the Boundaries," for a different view.

2. Parson, *Protecting the Ozone Layer*, 6.

3. Benedick, *Ozone Diplomacy*.

4. Parson, *Protecting the Ozone Layer*, chapter 3; Conway, *High-Speed Dreams*; Oreskes and Conway, *Merchants of Doubt*.

5. Stolarski and Cicerone, "Stratospheric Chlorine." Previous work by John Hampson (e.g., Hampson, "Chemiluminescent Emission Observed in the Stratosphere"), B. G. Hunt (e.g., Hunt, "Photochemistry of Ozone in a Moist Atmosphere"), and Paul Crutzen (e.g., Crutzen, "The Influence of Nitrogen Oxides") among others had explored other possible chain reactions involving oxides of hydrogen (HO_x) and oxides of nitrogen (NO_x) that could also contribute to ozone depletion. Their work paved the way for the chlorine work to follow.

6. Stolarski and Cicerone, "Stratospheric Chlorine," 1615; Molina and Rowland, "Stratospheric Sink for Chlorofluoromethanes."

7. Benedick, *Ozone Diplomacy*, 11; Roan, *Ozone Crisis*, 2.

8. Stolarski and Cicerone, "Stratospheric Chlorine," 1615.

9. Molina and Rowland, "Stratospheric Sink for Chlorofluoromethanes," 812.

10. US NAS, Panel on Atmospheric Chemistry, *Halocarbons*, 226.

11. Tony Cox, interview by Keynyn Brysse, Cambridge, UK, March 16, 2009.

12. Tony Cox, interview by Keynyn Brysse, Cambridge, UK, March 16, 2009.

13. Neil Harris, seminar/group interview by Keynyn Brysse, Cambridge, UK, March 16, 2009.

14. Rowland, "Chlorofluorocarbons and the Depletion of Stratospheric Ozone."

15. Farman, Gardiner, and Shanklin, "Large Losses of Total Ozone in Antarctica."

16. See, e.g., Christie, *The Ozone Layer*, chapter 7; and Parson, *Protecting the Ozone Layer*, chapter 6, for an overview; Callis and Natarajan, "Antarctic Ozone Minimum," for the solar theory; Mahlman and Fels, "Antarctic Ozone Decreases"; and Tung et al., "Antarctic Ozone Variations," for the dynamic theory; and Solomon et al., "On the Depletion of Antarctic Ozone"; and McElroy et al., "Reductions of Antarctic Ozone," for the chemical theory.

17. Halvorsen, "An Alarming and Mysterious 'Hole'"; NASA et al., "Antarctic Ozone: Initial Findings"; Roan, *Ozone Crisis*; Keith Shine, telephone interview by Keynyn Brysse, March 25, 2009.

18. NASA/WMO, *Atmospheric Ozone 1985*. Although "1985" is part of their title, the "Blue Books" were actually published in 1986.

19. Don Wuebbles, telephone interview by Keynyn Brysse, February 3, 2010.

20. Farman, Gardiner, and Shanklin, "Large Losses of Total Ozone in Antarctica."

21. NASA/WMO, *Atmospheric Ozone 1985*, 785.

22. This "inert" form is HCl and $ClNO_3$. In other words, these compounds, which do not react quickly in the gas phase and so were considered to be reservoirs of chlorine (and nitrogen), turn out to react very quickly on the surfaces of polar stratospheric cloud particles under a precise set of conditions often found in the spring Antarctic polar vortex (and not found in the nonpolar stratosphere).

23. Solomon, *The Coldest March*. Noctilucent clouds occur in the polar regions of the Northern Hemisphere as well.

24. F. Sherwood Rowland, interview by Keynyn Brysse, Irvine, CA, March 5, 2009.

25. Neil Harris, interview by Keynyn Brysse, Cambridge, UK, March 16, 2009.

26. Joseph Farman, interview by Keynyn Brysse, Cambridge, UK, March 16, 2009.

27. Crutzen is an atmospheric chemist who had earlier proposed a catalytic chain involving oxides of nitrogen (NO_x) that could cause ozone depletion. His predecessor at the Max Planck Institute for Chemistry in Mainz, Germany, was Christian Junge—the discoverer of the Junge layer—so Crutzen was institutionally and intellectually well positioned to consider the possible role of heterogeneous reactions in ozone depletion.

28. Cadle, Crutzen, and Ehhalt, "Heterogeneous Chemical Reactions in the Stratosphere."

29. Paul Crutzen, interview by Keynyn Brysse, Mainz, Germany, March 11, 2009.

30. Cadle, Crutzen, and Ehhalt, "Heterogeneous Chemical Reactions in the Stratosphere," 3384.

31. Martin, Judeikis, and Wun, "Heterogeneous Reactions of Cl and ClO in the Stratosphere," 5511. The argument is that gas-phase reaction of chlorine atoms with methane ties them up in the HCl reservoir. The aforementioned heterogeneous reaction of water and N_2O_5 creates a reservoir for nitrogen compounds that would likewise capture and remove Cl. Since the former reaction, thought to be faster than the latter, would dominate as a sink for Cl, the heterogeneous reaction could be safely ignored.

32. Sato and Rowland, "Current Issues in Our Understanding of the Stratosphere"; Rowland et al., "The Hydrolysis of Chlorine Nitrate."

33. F. Sherwood Rowland, interview by Keynyn Brysse, Irvine, CA, March 5, 2009.

34. Molina et al., "Upper Limit to the Rate of the Hydrogen Chloride + $ClONO_2$ Reaction," 3781.

35. Molina et al., "Antarctic Stratospheric Chemistry."

36. Glantz, Robinson, and Krenz, "Improving the Science of Climate-Related Impact Studies," 72.

37. Parson, *Protecting the Ozone Layer*, 28–29.

38. US NAS, Panel on Atmospheric Chemistry, *Halocarbons*. The NAS also produced an independent report to congress in 1972 on "the possible physical, biological, social and economic effects that might result from future aircraft operations in the stratosphere." US NAS, Climate Impact Committee, *Environmental Effects of Stratospheric Flight*, iv. This report was commissioned by the Department of Transportation in order to advise CIAP.

39. US NAS, Committee on Impacts of Stratospheric Change, *Halocarbons*.

40. US NAS, Committee on Impacts of Stratospheric Change, *Halocarbons*, vii–viii.

41. US NAS, Committee on Impacts of Stratospheric Change, *Halocarbons*, 12.

42. US NAS, Committee on Impacts of Stratospheric Change, *Halocarbons*, 12.

43. US NAS, Committee on Impacts of Stratospheric Change, *Halocarbons*, 15.

44. US NAS, Committee on Impacts of Stratospheric Change, *Halocarbons*, 7.

45. US NAS, Committee on Impacts of Stratospheric Change, *Halocarbons*, 8.

46. Quoted in Roan, *Ozone Crisis*, 81.

47. Roan, *Ozone Crisis*, 80.

48. Parson, *Protecting the Ozone Layer*, 238.

49. Quoted in Roan, *Ozone Crisis*, 80.

50. Parson, *Protecting the Ozone Layer*, 41. Despite Parson's claim that the academy report did more harm than good in the international arena, the Vienna Convention for the Protection of the Ozone Layer was adopted in 1985, having been negotiated since 1982.

51. Royal Commission on Environmental Pollution, *Pollution Control*. In its own words, "the RCEP is an independent standing body established in 1970 to advise the Queen, Government, Parliament, the devolved administrations and the public on environmental issues. RCEP website, https://www.preventionweb.net/organizations/6320.

52. UK DOE, Central Unit on Environmental Pollution, *Chlorofluorocarbons and Their Effect on Stratospheric Ozone*, 7.

53. UK DOE, Central Unit on Environmental Pollution, *Chlorofluorocarbons and Their Effect on Stratospheric Ozone*, 16.

54. UK DOE, Central Unit on Environmental Pollution, *Chlorofluorocarbons and Their Effect on Stratospheric Ozone*, iii.

55. UK DOE, Central Unit on Environmental Pollution, *Chlorofluorocarbons and Their Effect on Stratospheric Ozone*, iii.

56. UK DOE, Central Directorate on Environmental Pollution, *Chlorofluorocarbons and Their Effect on Stratospheric Ozone (Second Report)*, v.

57. UK DOE, Central Directorate on Environmental Pollution, *Chlorofluorocarbons and Their Effect on Stratospheric Ozone (Second Report)*, part 2, 202.

58. UK DOE, Central Directorate on Environmental Pollution, *Chlorofluorocarbons and Their Effect on Stratospheric Ozone (Second Report)*, part 2, 188.

59. UK DOE, Central Directorate on Environmental Pollution, *Chlorofluorocarbons and Their Effect on Stratospheric Ozone (Second Report)*, part 1, 5.

60. UK DOE, Central Directorate on Environmental Pollution, *Chlorofluorocarbons and Their Effect on Stratospheric Ozone (Second Report)*, part 1, 10.

61. Schefter, "New Aerodynamic Design, New Engines, Spawn a Revival of the SST."

62. Parson argues that the Fluorocarbon Panel exerted a significant influence on the reports produced by the Coordinating Committee on the Ozone Layer (CCOL), skewing them toward the CFC industry's antiregulation policy. Parson, *Protecting the Ozone Layer*, 100. Even where this was not happening, the very fact of the Fluorocarbon Panel's representation among CCOL members made some doubt the committee's objectivity. Parson claims that there were 14 countries represented at the first meeting (Parson, *Protecting the Ozone Layer*, 98), while Andersen, Sarma, and Sinclair say there were 16 countries represented. Andersen, Sarma, and Sinclair, *Protecting the Ozone Layer*, 48.

63. Robert T. Watson, interview by Keynyn Brysse, Norwich, UK, March 13, 2009. Parson regarded CCOL as a failure and claimed that it "failed because it lacked the resources or stature to articulate a broad scientific consensus, and because it had no policy body as its client." Parson, *Protecting the Ozone Layer*, 51. But not everyone who has written about CCOL has viewed it as a failure (e.g., Andersen, Sarma, and Sinclair, *Protecting the Ozone Layer*; Litfin, *Ozone Discourses*), and later international assessments done prior to the adoption of the Montreal Protocol had no more of a captive audience than did CCOL.

64. Robert T. Watson, interview by Keynyn Brysse, Norwich, UK, March 13, 2009.

65. CODATA members were drawn from the United Kingdom and Germany, as well as the United States. Their initial publication is Baulch et al., "Evaluated Kinetic and Photochemical Data."

66. Robert T. Watson, interview by Keynyn Brysse, Norwich, UK, March 13, 2009.

67. Robert T. Watson, interview by Keynyn Brysse, Norwich, UK, March 13, 2009.

68. Adrian Tuck, interview by Keynyn Brysse, Boulder, CO, June 24, 2009.

69. Robert T. Watson, interview by Keynyn Brysse, Norwich, UK, March 13, 2009.

70. As of this writing, Sundararaman is secretary of the IPCC.

71. NASA/WMO, *The Stratosphere 1981*, iii.

72. NASA/WMO, *The Stratosphere 1981*, iv. Parson attributes this comment to NASA's concern that the enhanced international participation in this assessment might "harm the scientific credibility of the report, by fostering the perception that scientific uncertainty or dissent might be suppressed in a politically motivated drive for consensus." Parson, *Protecting the Ozone Layer*, 102.

73. The Blue Books were "unprecedented both in the number of participating scientists and sponsoring organizations, and in [their] aim to provide comprehensive, authoritative reviews of every atmospheric topic relevant to stratospheric ozone." Parson, *Protecting the Ozone Layer*, 103.

74. NASA/WMO, *Atmospheric Ozone 1985*, 5; Parson, *Protecting the Ozone Layer*, 103.

75. This quotation is from an interview with Bojkov conducted in 1996, the transcript of which is part of the Edward A. Parson Stratospheric Ozone Collection, 1976–1997, Environmental Science and Public Policy Archives, box 1, folder 8, Harvard University.

76. NASA/WMO, *Atmospheric Ozone 1985*, 18.

77. Parson, *Protecting the Ozone Layer*, 105. Parson drew the general conclusion that a key feature of successful assessments is that they include "an international near-monopoly of relevant expertise to present an authoritative statement of present knowledge." Parson, *Protecting the Ozone Layer*, 267. While the truth of this claim depends on what one means by a "successful assessment," it may well fail as a sufficient condition even if it is necessary. Recent IPCC reports, which, if anything, are even more comprehensive and inclusive than the Blue Books and have articulated a clear consensus over a sustained period of time challenged by almost no one with appropriate scientific credentials and genuine scientific expertise, still are not universally accepted by stakeholders. The terms and conditions under which stakeholders accept or challenge scientific results depend on many more factors than the quality of those results and the manner in which they are gathered and presented.

78. NASA et al., *Report of the International Ozone Trends Panel 1988*.

79. Norman, "Satellite Data Indicate Ozone Depletion."

80. The diffuser plate is a ground aluminum plate used to reflect sunlight into the satellite instrument to measure solar radiance, used in calculating ozone levels. See NASA et al., *Report of the International Ozone Trends Panel 1988*, 1:207; and NASA, *Nimbus-7 Total Ozone Mapping Spectrometer (TOMS) Data Products User's Guide*, NASA Reference Publication 1323, November 1993, https://ntrs.nasa.gov/archive/nasa/casi.ntrs.nasa.gov/19940019882.pdf (accessed January 28, 2017).

81. The Dobson network consists of nearly 100 sites around the world that measure ozone using the Dobson spectrophotometer.

82. F. Sherwood Rowland, interview by Keynyn Brysse, Irvine, CA, March 5, 2009.

83. F. Sherwood Rowland, interview by Keynyn Brysse, Irvine, CA, March 5, 2009.

84. Robert T. Watson, interview by Keynyn Brysse, Norwich, UK, March 13, 2009.

85. NASA et al., *Report of the International Ozone Trends Panel 1988*, executive summary, 1–2; Edward Parson, "Stratospheric Ozone Collection," Harvard University, box 6, folder 61.

86. NASA et al., *Report of the International Ozone Trends Panel 1988*, executive summary, 2; Edward Parson "Stratospheric Ozone Collection," box 6, folder 61, Harvard University.

87. F. Sherwood Rowland, interview by Keynyn Brysse, Irvine, CA, March 5, 2009.

88. Robert T. Watson, interview by Keynyn Brysse, Norwich, UK, March 13, 2009.

89. Robert T. Watson, interview by Keynyn Brysse, Norwich, UK, March 13, 2009.

90. Scientists today frequently assert that to be credible, they (both individually and collectively) must remain studiously neutral on policy issues. The success of the Ozone Trends Panel calls that assumption at least somewhat into question. For further discussion, see chapter 5.

91. Heckert, letter of March 4, 1988, cited in Parson, *Protecting the Ozone Layer*, 156n36.

92. UN, "International Day for the Preservation of the Ozone Layer, 16 September," https://www.un.org/en/events/ozoneday/background.shtml (accessed December 29, 2017).

93. UNEP, *The Montreal Protocol on Substances That Deplete the Ozone Layer*, annex VI, http://ozone.unep.org/en/handbook-montreal-protocol-substances-deplete-ozone-layer/27506 (accessed October 12, 2016).

94. On the creation of the parties, see "Report of the Parties to the Montreal Protocol on the Work of Their First Meeting, Helsinki, May 2–5, 1989," https://ozone.unep.org/Meeting_Documents/mop/01mop/1mop-5e.shtm (accessed March 25, 2018); for the quotation regarding the executive summary, see UNEP, *The Montreal Protocol on Substances That Deplete the Ozone Layer*, annex VI, http://ozone.unep.org/en/handbook-montreal-protocol-substances-deplete-ozone-layer/27506 (accessed October 12, 2016).

95. NASA et al., *Scientific Assessment of Stratospheric Ozone*, vi.

96. Robert T. Watson, interview by Keynyn Brysse, Norwich, UK, March 13, 2009 (quote), and by telephone, September 28, 2011.

97. NASA et al., *Scientific Assessment of Stratospheric Ozone*. The full 1989 assessment comprised two volumes: the first was the scientific report chaired by Watson and Albritton; volume 2 was a separate *Appendix: AFEAS Report* (the Alternative Fluorocarbon Environmental Acceptability Study), prepared by 15 CFC-producing companies from around the world. As the preface to the first volume explains, the AFEAS document was highly useful to the scientific assessment authors in preparing their report (chiefly in preparing chapter 4, "Halocarbon Ozone Depletion and Global Warming Potentials"), and so it was decided to include AFEAS as part of the Montreal Protocol–mandated assessment. For reasons of space, however, and because we are focusing on science, not industry, we will not discuss AFEAS further, and when we refer to the 1989 assessment, unless otherwise indicated, we shall be referring only to the first volume, the scientific report.

98. NASA et al., *Scientific Assessment of Ozone Depletion: 1991*; NOAA et al., *Scientific Assessment of Ozone Depletion: 1994*; NOAA et al., *Scientific Assessment of Ozone Depletion: 1998*; NOAA et al., *Scientific Assessment of Ozone Depletion: 2002*; NOAA et al., *Scientific Assessment of Ozone Depletion: 2006*; NOAA et al., *Scientific Assessment of Ozone Depletion: 2010*; NOAA et al., *Scientific Assessment of Ozone Depletion: 2014*. At the time of writing, the *Scientific Assessment of Ozone Depletion: 2018* is in preparation. Watson and Albritton were cochairs (along with various others) on all of them up to and including the 2006 assessment.

99. David Fahey, interview by Keynyn Brysse, Boulder, CO, June 24, 2009.

100. Robert T. Watson, interview by Keynyn Brysse, Norwich, UK, March 13, 2009.

101. Chris Mooney, "Some Like It Hot," in Special Reports: As the World Burns, *Mother Jones* 30, no. 3 (May–June 2005): 36–94; Oreskes and Conway, *Merchants of Doubt*.

102. Robert T. Watson, interview by Keynyn Brysse, Norwich, UK, March 13, 2009.

103. John Pyle, interview by Keynyn Brysse, Cambridge, UK, March 17, 2009.

104. Robert T. Watson, interview by Keynyn Brysse, Norwich, UK, March 13, 2009.

105. Mack McFarland, interview by Keynyn Brysse, Wilmington, DE, July 31, 2009.

106. Robert T. Watson, interview by Keynyn Brysse, Norwich, UK, March 13, 2009.

107. John Pyle, interview by Keynyn Brysse, Cambridge, UK, March 17, 2009.

108. John Pyle, interview by Keynyn Brysse, Cambridge, UK, March 17, 2009.

109. Robert T. Watson, interview by Keynyn Brysse, Norwich, UK, March 13, 2009.

110. Guy Brasseur, interview by Keynyn Brysse, Boulder, CO, June 25, 2009.

111. Neil Harris, interview by Keynyn Brysse, Cambridge, UK, March 16, 2009.

112. John Pyle, interview by Keynyn Brysse, Cambridge, UK, March 17, 2009.

113. Tony Cox, seminar/group interview by Keynyn Brysse, Cambridge UK, March 16, 2009.

114. Robert T. Watson, interview by Keynyn Brysse, Norwich, UK, March 13, 2009.

115. Brasseur is the current director of the Climate Service Center, Germany, and former associate director of the US National Center for Atmospheric Research (NCAR). Dr. Brasseur's PhD, awarded in 1974, examined the effect of NO_x on stratospheric ozone, specifically in the context of the proposed SSTs. He then did a stint in democratic politics, serving from 1977 to 1981 as an elected member of the Belgian house of representatives, and was a delegate to the Parliamentary Assemblies of the Council of Europe (Strasbourg, France) and of the Western European Union (Paris, France) before returning to scientific research. In 1984, Brasseur made a five-month visit to the Max Planck Institute for Chemistry in Mainz, where he worked with Paul Crutzen, before moving to NCAR. See "Guy Brasseur—Summary," http://www.mpimet.mpg.de/en/staff/ohneabteilung/guy-brasseur/ (accessed October 12, 2016).

116. Robert T. Watson, interview by Keynyn Brysse, Norwich, UK, March 13, 2009.

117. Parson concurs with Watson's judgment. Parson, *Protecting the Ozone Layer*, 107–108. In this particular case, however, Watson's concern may have been unwarranted, as the resolution of the European Parliamentary Assembly that resulted from Brasseur's study supported action on ozone depletion. On the other hand, in comparison to the United States and Scandinavia, which had already enacted bans on CFCs in aerosols two years before, the European position was a weak one.

118. Indeed, this is the heart of the strategy of "merchandising doubt." Contrarians try to challenge or deny scientific consensus in order to argue that the knowledge is not stable and therefore does not provide a good basis for decision making. See Oreskes and Conway, *Merchants of Doubt*.

119. Recall this is before the ozone hole had been discovered and before its explanation through heterogeneous chemistry.

120. Guy Brasseur, email to Keynyn Brysse, September 25, 2013; see also Counsel of Europe / Conseil de L'Europe Resolution 733 (1980).

121. Robert T. Watson, interview by Keynyn Brysse, Norwich, UK, March 13, 2009.

122. Robert T. Watson, interview by Keynyn Brysse, Norwich, UK, March 13, 2009.

123. John Pyle, interview by Keynyn Brysse, Cambridge, UK, March 17, 2009.

124. Robert T. Watson, interview by Keynyn Brysse, Norwich, UK, March 13, 2009. No one knows who invented the elevator pitch, but evidence indicates it predates Bob Watson.

125. O. Brian Toon, interview by Keynyn Brysse, Boulder, CO, June 25, 2009. Moreover, climate change was already established in 2009.

126. Michael McIntyre, seminar/group interview by Keynyn Brysse, Cambridge UK, March 16, 2009.

127. Lambright, *NASA and the Environment*, 22.

128. Albritton, "What Should Be Done in a Science Assessment?," 69; see also the discussion in Cash and Clark, *From Science to Policy*, 13.

129. NOAA et al., *Scientific Assessment of Ozone Depletion: 1994*, section 13.3.

130. Wuebbles, "Chlorocarbon Emission Scenarios"; NASA et al., *Scientific Assessment of Stratospheric Ozone*.

131. Hoffman and Gibbs, *Future Concentrations of Stratospheric Chlorine and Bromine*.

132. Parson, *Protecting the Ozone Layer*, 160.

133. Daniel, Solomon, and Albritton, "On the Evaluation of Halocarbon Radiative Forcing."

134. Robert T. Watson, interview by Keynyn Brysse, Norwich, UK, March 13, 2009.

135. Tony Cox, interview by Keynyn Brysse, Cambridge, UK, March 16, 2009.

136. Robert T. Watson, interview by Keynyn Brysse, Norwich, UK, March 13, 2009.

137. Albritton, "What Should Be Done in a Science Assessment?," 70.

138. Don Wuebbles, telephone interview by Keynyn Brysse, February 3, 2010.

139. Jonathan Shanklin and Michael McIntyre, seminar/group interview by Keynyn Brysse, Cambridge UK, March 16, 2009.

140. For the impact of simplifying assumptions in the debate over continental drift, see Oreskes, *The Rejection of Continental Drift*.

CHAPTER FOUR

1. Justin Gillis and Kenneth Chang, "Scientists Warn of Rising Oceans from Polar Melt," *New York Times*, May 12, 2014, https://www.nytimes.com/2014/05/13/science/earth/collapse-of-parts-of-west-antarctica-ice-sheet-has-begun-scientists-say.html (accessed January 18, 2017).

2. There are three principal causes of global sea level rise in a warming world: retreat of the Greenland and Antarctic Ice Sheets; melting of mountain glaciers; and thermal expansion of warmed sea water (which causes it to take up more volume). At present, these three sources are making about equal contributions to measured sea level rise, but the ice sheet contribution has the greatest potential for rapid and dramatic increase and therefore potentially catastrophic consequences.

3. O'Reilly, Oreskes, and Oppenheimer, "Rapid Disintegration of Projections."

4. Brysse et al., "Climate Change Prediction."

5. Aronova, Baker, and Oreskes, "Big Science and Big Data in Biology."

6. The USSR constructed its own station near the Southern Pole of Inaccessibility on Lake Vostok (the most inland Antarctic location) and small bases within each of the territorial claims of other countries.

7. Thomas, "Research Agendas in Climate Studies."

8. Naylor, Dean, and Siegert, "The IGY and the Ice Sheet." These traverses were inspired by the earlier work of the Norwegian-British-Swedish expedition of 1949–1952, which explored the participating states' Antarctic claims and carried out cooperative international science and logistical work such as mapping unknown space and identifying crevasses. Once the Antarctic Treaty (1959) put territorial claims on ice, states maintained sovereignty (and symbolically continued to fight the Cold War) by demonstrating their long-term, comprehensive scientific activities in the Antarctic (see figure 4.1).

9. Naylor, Dean, and Siegert, "The IGY and the Ice Sheet."

10. The renaming of the Ross Sea sector ice streams from alphabetical indicators to names of renowned American scientists is well accepted among US scientists but remains controversial among scientists from other countries, who also conducted decades of research in the area.

11. Hermann Engelhardt, interview by Jessica O'Reilly, Pasadena, CA, April 23, 2009.

12. Weart, *Discovery of Global Warming*.

13. Mercer, "West Antarctic Ice Sheet and CO_2 Greenhouse Effect."

14. In later years Hughes would challenge the assumption that climate change would be detrimental; see Climate Depot, "Prominent Scientist Dissents," http://www.climatedepot.com/2015/01/08/prominent-scientist-dissents-internationally-renowned-glaciologist-declares-global-warming-is-good-not-bad-its-going-to-be-a-big-plus/ (accessed December 29, 2017).

15. AAAS, *Conference Proceedings*.

16. Munk and Revelle, "Geophysical Interpretation of Irregularities"; Munk and Revelle, "Sea Level and the Rotation of the Earth"; Revelle and Suess, "Carbon Dioxide Exchange between Atmosphere and Ocean."

17. Rainger, "Constructing a Landscape for Postwar Science"; Oreskes, *Science on a Mission*.

18. AAAS, *Conference Proceedings*, 14–15. Revelle also participated in the establishment of the Scientific Committee on Antarctic Research (SCAR) in its formative, post-IGY years, 1958–1960. SCAR continues to be the international organizational scientific body for Antarctic research, putting together workshops so that international research teams can coordinate research, hosting a massive annual research conference, and providing scientific advice on management decisions at the Antarctic Treaty Consultative Meetings. During the post-IGY era, SCAR committed itself to carrying forward some of the key research missions conducted during IGY, including the continued study of the West Antarctic Ice Sheet.

19. AAAS, *Conference Proceedings*, 5.

20. AAAS, *Conference Proceedings*, 6–7.

21. AAAS, *Conference Proceedings*, 14.

22. AAAS, *Conference Proceedings*, 14.

23. AAAS, *Conference Proceedings*, 17–18.

24. AAAS, *Conference Proceedings*, 79–80.

25. AAAS, *Conference Proceedings*, 91–92.

26. AAAS, *Conference Proceedings*, 101.

27. AAAS, *Conference Proceedings*, 103–104. Two processes control disintegration of land ice into sea water: direct melting, which is especially prominent in Greenland and for mountain glaciers, and dynamic flow of the ice toward and then into the sea in the form of floating ice shelves that are still attached to the land-based or "grounded" ice. Eventually, ice shelves fragment into free-floating icebergs. Flow of ice into the sea is the dominant process by which WAIS loses ice and contributes to sea level rise. The grounding line is the boundary between grounded ice and floating ice shelves.

28. AAAS, *Conference Proceedings*, 108.

29. AAAS, *Conference Proceedings*, 163.

30. AAAS, *Conference Proceedings*, 217–218.

31. AAAS, *Conference Proceedings*, 272–275.

32. AAAS, *Conference Proceedings*, 278–279.

33. AAAS, *Conference Proceedings*, 285–286.

34. AAAS, *Conference Proceedings*, 365.

35. AAAS, *Conference Proceedings*, 533.

36. CDAC, NRC, *Increasing Carbon Dioxide and the West Antarctic Ice Sheet*, 1982, appendix B.

37. IEEE Power Engineering Society, "Carbon Dioxide Proliferation," 21.

38. IEEE Power Engineering Society, "Carbon Dioxide Proliferation," 22.

39. IEEE Power Engineering Society, "Carbon Dioxide Proliferation," 26.

40. IEEE Power Engineering Society, "Carbon Dioxide Proliferation," 30.

41. Revelle, "Carbon Dioxide and World Climate," 35.

42. Almost all of the materials used in this section are from the Roger Randall Dougan Revelle (1909–1991) Papers, housed in the Scripps Institute of Oceanography archives at the University of California, San Diego. Nierenberg was skeptical that climate change was a serious issue, and Revelle knew this, so it may be that Revelle organized the workshop in order to make sure that his chapter on sea level rise was informed by the most up-to-date science available.

43. CDAC, NRC, *Increasing Carbon Dioxide and the West Antarctic Ice Sheet*, 1.

44. CDAC, NRC, *Increasing Carbon Dioxide and the West Antarctic Ice Sheet*, 3.

45. CDAC, NRC, *Increasing Carbon Dioxide and the West Antarctic Ice Sheet*, 3.

46. CDAC, NRC, *Increasing Carbon Dioxide and the West Antarctic Ice Sheet*, 3.

47. CDAC, NRC, *Increasing Carbon Dioxide and the West Antarctic Ice Sheet*, 4.

48. The Eemian interglacial took place 130,000–114,000 BP.

49. CDAC, NRC, *Increasing Carbon Dioxide and the West Antarctic Ice Sheet*, 4.

50. Gornitz, Lebedeff, and Hansen, "Global Sea Level Trend in the Past Century."

51. CDAC, NRC, *Increasing Carbon Dioxide and the West Antarctic Ice Sheet*, 5.

52. CDAC, NRC, *Increasing Carbon Dioxide and the West Antarctic Ice Sheet*, 5.

53. CDAC, NRC, *Increasing Carbon Dioxide and the West Antarctic Ice Sheet*, 5.

54. CDAC, NRC, *Increasing Carbon Dioxide and the West Antarctic Ice Sheet*, 6, emphasis in original.

55. CDAC, NRC, *Increasing Carbon Dioxide and the West Antarctic Ice Sheet*, 7.

56. CDAC, NRC, *Increasing Carbon Dioxide and the West Antarctic Ice Sheet*, 7.

57. CDAC, NRC, *Increasing Carbon Dioxide and the West Antarctic Ice Sheet*, 7, emphasis in original.

58. CDAC, NRC, *Increasing Carbon Dioxide and the West Antarctic Ice Sheet*, 7.

59. CDAC, NRC, *Increasing Carbon Dioxide and the West Antarctic Ice Sheet*, 2.

60. Institute for Energy Analysis, *Proceedings: Carbon Dioxide Research Conference*.

61. Bentley, "West Antarctic Ice Sheet."

62. Bentley, "West Antarctic Ice Sheet," 44.

63. DeAngelis and Skvarca, "Glacial Surge after Ice Sheet Collapse"; Rignot et al., "Accelerated Ice Discharge from the Antarctic Peninsula."

64. Part of the difference of views may have originated in Hughes's research focus on the Amundsen Sea region of WAIS, where changes were occurring faster than elsewhere, while Bentley focused on the apparently more stable region draining into the Ross Ice Shelf.

65. Gordon, "Comments about the Ocean Role in the Antarctic Glacial Balance."

66. CDAC, NRC, *Changing Climate*, 433.

67. CDAC, NRC, *Changing Climate*, 433.

68. CDAC, NRC, *Changing Climate*, 435.

69. CDAC, NRC, *Changing Climate*, 436.

70. CDAC, NRC, *Changing Climate*, 435.

71. CDAC, NRC, *Changing Climate*, 442.

72. CDAC, NRC, *Changing Climate*, 444.

73. CDAC, NRC, *Changing Climate*, 444.

74. CDAC, NRC, *Changing Climate*, 444.

75. CDAC, NRC, *Changing Climate*, 446.

76. CDAC, NRC, *Changing Climate*, 446.

77. US EPA, *Projecting Future Sea Level Rise*, 22–23.

78. US EPA, *Projecting Future Sea Level Rise*, 22–24.

79. US EPA, *Projecting Future Sea Level Rise*, 24.

80. US EPA, *Projecting Future Sea Level Rise*, 1983, 31.

81. US EPA, *Projecting Future Sea Level Rise*, 33.

82. US EPA, *Projecting Future Sea Level Rise*, 33–34.

83. US EPA, *Projecting Future Sea Level Rise*, vi.

84. US EPA, *Projecting Future Sea Level Rise*, 33.

85. US EPA, *Projecting Future Sea Level Rise*, 29.

86. Oreskes, Conway, and Shindell, "From Chicken Little to Dr. Pangloss."

87. NRC, *Glaciers, Ice Sheets, and Sea Level*, xi.

88. NRC, *Glaciers, Ice Sheets, and Sea Level*, 1–2.

89. NRC, *Glaciers, Ice Sheets, and Sea Level*, 30–31.

90. NRC, *Glaciers, Ice Sheets, and Sea Level*, 31.

91. NRC, *Glaciers, Ice Sheets, and Sea Level*, 31.

92. NRC, *Glaciers, Ice Sheets, and Sea Level*, 33.

93. NRC, *Glaciers, Ice Sheets, and Sea Level,* 33–34.

94. NRC, *Glaciers, Ice Sheets, and Sea Level,* 47.

95. Budd, Jenssen, and Smith, "Three-Dimensional Time-Dependent Model."

96. MacAyeal and Thomas, "Numerical Modeling of Ice Shelf Motion"; Fastook, "West Antarctica, the Sea-Level Controlled Marine Instability"; Lingle, "Numerical Model of Interactions between a Polar Ice Stream and the Ocean."

97. NRC, *Glaciers, Ice Sheets, and Sea Level,* 58, emphasis in original.

98. NRC, *Glaciers, Ice Sheets, and Sea Level,* 65.

99. NRC, *Glaciers, Ice Sheets, and Sea Level,* 65–67.

100. Though Chamberlin did indeed write about a potential relationship between carbon dioxide and atmospheric warming, he did not "sound an alarm" about the potential implications that could come from it. His research was concerned with past climate events, not future ones. Fleming, *Historical Perspectives on Climate Change.*

101. IIASA, *Life on a Warmer Earth,* v.

102. IIASA, *Life on a Warmer Earth,* 3.

103. IIASA, *Life on a Warmer Earth,* 15.

104. IIASA, *Life on a Warmer Earth,* 19–20.

105. IIASA, *Life on a Warmer Earth,* 54–44.

106. I. Smith, *Carbon Dioxide—Emissions and Effects.*

107. International Energy Agency (IEA) Clean Coal Centre, http://www.iea-coal.org.uk (accessed November 4, 2016). According to its website, the United Kingdom–based IEA Clean Coal Centre is "the world's foremost provider of information on the clean and efficient use of coal worldwide, particularly clean coal technologies, in a balanced and objective way, without political or commercial bias." http://www.iea-coal.org.uk/site/2010/home (accessed December 29, 2017).

108. I. Smith, *Carbon Dioxide—Emissions and Effects,* 96.

109. I. Smith, *Carbon Dioxide—Emissions and Effects,* 68–71.

110. I. Smith, *Carbon Dioxide—Emissions and Effects,* 95.

111. "Working Group I: The Scientific Basis," executive summary, https://www.ipcc.ch/ipccreports/tar/wg1/440.htm (accessed December 29, 2017); IPCC, *Climate Change 2007,* 5.

112. Bolin, *History of the Science and Politics of Climate Change*; Hulme, *Why We Disagree about Climate Change*; S. Schneider, *Science as a Contact Sport.*

113. IPCC, *Climate Change: The IPCC Scientific Assessment.*

114. Johannes Oerlemans, interview by Jessica O'Reilly, Utrecht, the Netherlands, July 6, 2009.

115. Johannes Oerlemans, interview by Jessica O'Reilly, Utrecht, the Netherlands, July 6, 2009.

116. Johannes Oerlemans, interview by Jessica O'Reilly, Utrecht, the Netherlands, July 6, 2009.

117. That scientists from the United States did not participate in a representative way in the first IPCC report is something often repeated but poorly explained. Interview participants have suggested that the IPCC was seen as just another in a large series of assessments and that it felt "more European" at its outset.

118. Johannes Oerlemans, interview by Jessica O'Reilly, Utrecht, the Netherlands, July 6, 2009.

119. IPCC, *Climate Change: The IPCC Scientific Assessment,* 261.

120. IPCC, *Climate Change: The IPCC Scientific Assessment,* 261.

121. IPCC, *Climate Change: The IPCC Scientific Assessment,* xxx.

122. IPCC assessments are written by three working groups. Working Group I is concerned with the physical science of climate change, including ice and sea level rise. Each working group produces a Summary for Policy Makers.

123. IPCC, *Climate Change 1995,* 4.

124. IPCC, *Climate Change 1995,* 6.

125. IPCC, *Climate Change 1995,* 363.

126. IPCC, *Climate Change 1995,* 364.

127. IPCC, *Climate Change 1995,* 389.

128. IPCC, *Climate Change 1995*, 389.

129. The Synthesis Report is meant to state the most policy-important messages of all three working group reports.

130. IPCC, *Climate Change 2001*, 9.

131. IPCC, *Climate Change 2001*, 15.

132. IPCC, *Climate Change 2001*, 14.

133. IPCC, *Climate Change 2001*, 14, emphasis added.

134. IPCC, *Climate Change 2001*, 16.

135. IPCC, *Climate Change 2001*, 16.

136. IPCC, *Climate Change 2001*, 17.

137. Oppenheimer et al., "Climate Change"; Rahmstorf, "Semi-Empirical Approach to Projecting Future Sea-level Rise."

138. Rignot et al., "Recent Antarctic Ice Mass Loss from Radar Interferometry"; Chen et al., "Accelerated Antarctic Ice Loss from Satellite Gravity Measurements."

139. Velicogna, "Increasing Rates of Ice Mass Loss from the Greenland and Antarctic Ice Sheets."

140. Rahmstorf, "Semi-Empirical Approach to Projecting Future Sea-level Rise"; Vermeer and Rahmstorf, "Global Sea Level Linked to Global Temperature."

141. O'Reilly, Oreskes, and Oppenheimer, "Rapid Disintegration of Projections."

142. IPCC, *Climate Change 2007*, 14.

143. IPCC, *Climate Change 2007*, 17.

144. IPCC, *Climate Change 2007*, 340. The italicized phrases in this and subsequent excerpts are technical terms used by the IPCC to characterize numerical probabilities of climate events.

145. IPCC, *Climate Change 2007*, 340.

146. IPCC, *Climate Change 2007*, 342.

147. The table appeared originally in chapter 5, without the caveat; IPCC, *Climate Change 2007*, 419. It was reproduced (with added caveat) in the Synthesis Report.

148. IPCC, *Climate Change 2007*, 8.

149. IPCC, *Climate Change 2007*, 12. One of the authors of this volume, Michael Oppenheimer, was a lead author of chapter 19 of the Working Group II report of AR4, contributing to the language in that chapter, which formed part of the basis for this statement in the Synthesis Report.

150. IPCC, *Climate Change 2007*, 20. Original emphasis.

151. IPCC, *Climate Change 2007*, 5.

152. Brysse et al., "Climate Change Prediction."

153. Oreskes and Belitz, "Philosophical Issues in Model Assessment."

154. Rahmstorf, "Semi-Empirical Approach to Projecting Future Sea-Level Rise."

155. Huybrechts, "3-D Model for the Antarctic Ice Sheet"; Huybrechts and Oerlemans, "Response of the Antarctic Ice Sheet to Future Greenhouse Warming."

156. Weertman, "Stability of the Junction of an Ice Sheet and an Ice Shelf;" Patt, "Extreme Outcomes."

CHAPTER FIVE

1. Proctor, *Value-Free Science?*, Putnam, *Collapse of the Fact/Value Dichotomy*; Kincaid, Dupré, and Wylie, *Value-Free Science?* For discussion of many of the themes of this chapter, see Jamieson, *Reason in a Dark Time*, chapter 3.

2. Acid Precipitation Act of 1980, Public Law 96-294, *US Statutes at Large* 94 (1980), 771. See also NRC, US National Committee for CODATA, "National Acid Precipitation Assessment Program."

3. IPCC, "Statement on IPCC Principles and Procedures," http://www.ipcc.ch/pdf/press/ipcc-statement-principles-procedures-02-2010.pdf (accessed January 5, 2017).

4. UNFCCC, Article 2, https://unfccc.int/resource/docs/convkp/conveng.pdf (accessed January 5, 2017).

5. Article 2 of the 1992 UNFCCC states the treaty's long-term objective: stabilizing "greenhouse gas concentrations in the atmosphere at a level that would prevent dangerous anthropogenic interference with the climate system." In 2009, the parties to the convention drew on scientific evidence, diverse cultural views of the importance of various climate impacts on different populations worldwide, previous international statements and commitments, and to some extent the possibility of achieving such a goal to define "dangerous interference" as a warming of 2°C above the preindustrial (circa 1750) global mean temperature. https://unfccc.int/resource/docs/convkp/conveng.pdf (accessed January 21, 2017). For the text of the Paris Agreement, see http://unfccc.int/paris_agreement/items/9485.php (accessed January 5, 2017).

6. On current developments vis-à-vis the West Antarctic Ice Sheet, see, for example, http://www.nasa.gov/jpl/news/antarctic-ice-sheet-20140512/#.WBDUno9Viko (accessed January 5, 2017).

7. Keller, "What Are Climate Scientists to Do?"; see also Oreskes, "How Earth Science Became a Social Science." Climate skeptics, contrarians, and deniers cite this as a reason to reject the IPCC findings. They do not accept the premise that 2°C of climate warming is dangerous or otherwise problematic. See, for example, Lomborg, *Skeptical Environmentalist*; see also Oreskes, "My Facts Are Better Than Your Facts."

8. Jasanoff, *Designs on Nature*; Latour, *Politics of Nature*; Latour, *We Have Never Been Modern*; Kitcher, *Science, Truth, and Democracy*; Miller and Edwards, *Changing the Atmosphere*; Hilgartner, Miller, and Hagendijk, *Science and Democracy*.

9. On matters of concern, see Latour, "Why Has Critique Run Out of Steam?" Bjorn Lomborg (in *Skeptical Environmentalist*) has argued that climate change is not as serious a matter as many climate scientists hold it to be and that it is overblown compared to the problems of hunger and poverty. Pope Francis explicitly addressed that dichotomy by linking climate change and poverty: see Pope Francis, *Encyclical Letter: Laudato Si' of the Holy Father Francis on Care for Our Common Home*, https://laudatosi.com/watch (accessed January 5, 2017). For discussion, see Jamieson, "Theology and Politics in *Laudato Si'*."

10. Scientists have not only spoken up about technological risks after the fact: in the 1970s, molecular biologists organized to discuss and control potential risks from recombinant DNA research and to recommend policies for how such research should be conducted. See Berg and Singer, "Recombinant DNA Controversy." In the 1990s, biologists issued the "Wingspread Statement on the Precautionary Principle," which expressed support for the application of the precautionary principle with respect to the problem of endocrine-disrupting chemicals: Science & Environmental Health Network, "Wingspread Conference on the Precautionary Principle," http://www.sehn.org/wing.html (accessed January 5, 2017). These activities can be distinguished from formal policy recommendations made in the context of an institutionalized assessment.

11. There is a large literature on Bohr: a good starting place regarding his engagement on implications of nuclear warfare and the arms race is Aaserud, "The Scientist and the Statesmen"; see also Ole Wæver, "Open Science to Fight Big Threats Like the Nuclear Bomb," *The Conversation*, January 2, 2014. https://theconversation.com/open-science-to-fight-big-threats-like-the-nuclear-bomb-21591 (accessed January 5, 2017). On scientists coming under suspicion for their political views, real or imagined, see J. Wang, *American Science in an Age of Anxiety*; and Oreskes and Rainger, "Science and Security before the Atomic Bomb."

12. See James Risen, "FBI Clears Top Physicists of Passing A-Bomb Secrets—Weapons: Allegations in Ex-KGB Officer's Book that Oppenheimer, Bohr, Fermi and Szilard had Given Postwar Aid to Soviets Provoked Outrage," *Los Angeles Times*, May 2, 1995, http://articles.latimes.com/1995

-05-02/news/mn-61373_1_atomic-bomb (accessed January 5, 2017). On scientists as uncertain allies, see Oreskes and Rainger, "Science and Security before the Atomic Bomb."

13. Schweber, *Nuclear Forces*.

14. On the role of the Joliot-Curies in France, see Weart, *Scientists in Power*. On Hans Bethe, see Schweber, *Nuclear Forces*.

15. Rhodes, *Making of the Atomic Bomb*, 697, 749–750; Thorpe, *Oppenheimer*, 7, 123, 156. See also Sherwin, *A World Destroyed*, 217–219; Alperovitz, *Decision to Use the Atomic Bomb*, 190, 604–607; J. Franck, D. Hughes, L. Szilard, T. Hogness, E. Rabinowitch, G. Seaborg, and C. J. Nickson, "The 'Franck Report': A Report to the Secretary of War, June 1945," http://www.fas.org/sgp/eprint/franck.html (accessed January 5, 2017).

16. Z. Wang, *In Sputnik's Shadow*, 19–20.

17. Federation of American Scientists, *Public Interest Report: 70 Years of Science Serving Society and Counting*, https://fas.org/wp-content/uploads/2016/05/70thannivpir_finaldigital.pdf (accessed January 5, 2017); see also A. Smith, *A Peril and a Hope*; Rhodes, *Making of the Atomic Bomb*.

18. Goldberg, "Creating a Climate of Opinion"; Rhodes, *Making of the Atomic Bomb*; Rhodes, *Dark Sun*; Sherwin, *A World Destroyed*.

19. Thorpe, *Oppenheimer*, 191, 286. Oppenheimer was inconsistent. After the war, he spoke to many issues outside the narrowly technical, including, famously, arguing that physicists, having built the bomb, now knew "sin." Later he would claim that the sin he referred to was the sin of pride, but that was not how it was generally interpreted it at the time. Most saw it as suggesting that scientists bore responsibility for what they had done and thus for thinking as well about its future and control. See also Schweber, *In the Shadow of the Bomb*.

20. Z. Wang, *In Sputnik's Shadow*, 22. Michael Gordin (personal communication, April 2014) notes that there appears to be some difference related to national origin and education: European-born scientists like Bethe, Szilard, and Bohr saw themselves as broadly trained intellectuals for whom it was appropriate to consider and speak on the ethical aspects of science; Feynman was perhaps not atypical of American-born scientists, who saw their role more narrowly.

21. Sherwin, *A World Destroyed*. See also Rhodes, *Making of the Atomic Bomb*; and Alperovitz, *Decision to Use the Atomic Bomb*.

22. Z. Wang, *In Sputnik's Shadow*, 20.

23. US Department of State, Office of the Historian, "The Acheson-Lilienthal & Baruch Plans, 1946," https://history.state.gov/milestones/1945-1952/baruch-plans (accessed January 5, 2017).

24. Hewlett and Duncan, *Atomic Shield*; York, *Advisors*; Z. Wang, "Scientists and Arms Control."

25. Quoted in Bernstein and Galison, "In Any Light," 269.

26. The hearing board that took that action against him cited, as one reason, the fact that he had inappropriately strayed beyond the technical and into the moral and political realm. Z. Wang, *In Sputnik's Shadow*, 46.

27. Historians Martin Sherwin and Kai Bird suggest that American scientists took this to heart and now believed that they could serve the state "only as experts on narrow scientific issues." Bird and Sherwin, *American Prometheus*, 549.

28. Z. Wang, *In Sputnik's Shadow*. The Jason committee was the idea of scientists who felt they had important advice to offer the Department of Defense and needed a recognized channel for doing so. Finkbeiner, *The Jasons*; Marvin Goldberger, personal communication.

29. For a comparative study of science policy in the United States, Germany, and the United Kingdom, see Jasanoff, *Designs on Nature*. John Krige has documented how both American scientists and political leaders during the Cold War tended to assume US superiority in all things scientific and technical and therefore to doubt that they had much to learn from their European counterparts. Krige, *American Hegemony and the Postwar Reconstruction of Europe*; Krige, *Sharing Knowledge, Shaping Europe*.

30. Attributed to Winston Churchill, quoted in Randolph S. Churchill, *Twenty-One Years*, 127. See also Z. Wang, *In Sputnik's Shadow*, 2.

31. Z. Wang, *In Sputnik's Shadow*, 64.

32. Z. Wang, *In Sputnik's Shadow*, 64.

33. The claim that the Soviets were ahead of the United States in nuclear capacity. Similar claims would be made throughout the Cold War regarding various Soviet capacities; later, most of these claims were shown to have been incorrect, or at least exaggerated. See Oreskes and Conway, *Merchants of Doubt*, chapter 2.

34. Quoted in Z. Wang, *In Sputnik's Shadow*, 110.

35. Quoted in Z. Wang, *In Sputnik's Shadow*, 104.

36. Z. Wang, *In Sputnik's Shadow*, 104. Oreskes was a colleague of Herbert York at the University of California, San Diego (UCSD) for many years, during which he regaled UCSD undergraduates with stories of President Eisenhower. It seems that they had a close relationship of mutual respect.

37. Z. Wang, *In Sputnik's Shadow*, 2.

38. Historian Zuoyue Wang writes, "Crucially Eisenhower agreed with PSAC on the need for science advising to integrate technical evaluations and policy considerations." Z. Wang, *In Sputnik's Shadow*, 3.

39. Quoted in Z. Wang, *In Sputnik's Shadow*, 14.

40. Quoted in Z. Wang, *In Sputnik's Shadow*, 244.

41. Z. Wang, *In Sputnik's Shadow*, 9.

42. On the Jasons see Finkbeiner, *The Jasons*; and Marvin Goldberger, "Science, Policy and Conflict," guest lecture at UCSD, January 24, 2008.

43. Zuoyue Wang suggests that Johnson's decision to develop the Sentinel "thin ABM" system over scientists' opposition, and particularly Robert McNamara's false suggestions in public that scientists had endorsed this approach, led PSAC scientists for the first time to begin to dissent publicly. Thus the ABM "split" began in the Johnson years but culminated under Nixon. See Z. Wang, *In Sputnik's Shadow*, 280.

44. Carson, *Silent Spring*.

45. Commoner, *Closing Circle*; Ehrlich, *Population Bomb*; see also Sabin, *The Bet*.

46. Commoner noted in the closing pages of *The Closing Circle* that in "our progress-minded society, anyone who presumes to explain a serious problem is expected to solve it as well" (299); Hardin, Ehrlich, and Carson made forays into suggesting possible solutions to the problems they had highlighted. Commoner attempted to demur other than to suggest that the problem of environmental destruction was a very large one, linked to our way of life and forms of governance. However, he did express the view that those, like Hardin, who succumbed to extreme conclusions about population control suffered from a failure of imagination, accepting "as unchangeable the present governance of a social good (the commons, or the ecosphere) by private need" (295). Later, however, he became a vocal advocate of renewable energy, founded the Citizens Party in 1980, and ran for president on that party's ticket. See Jeff Faux, "Barry Commoner and the Dream of a Liberal Third Party," *American Prospect*, October 9, 2012, http://prospect.org/article/barry-commoner-and-dream-liberal-third-party (accessed January 1, 2018).

47. Z. Wang, *In Sputnik's Shadow*, 303.

48. It might similarly be argued that that advice supplied by a national assessment such as NAPAP would be viewed as more objective than the advice offered by individual scientists serving at the discretion of a particular president.

49. The 1983 National Research Council (NRC) "red book," *Risk Assessment in the Federal Government: Managing the Process*—in effect an assessment of risk assessment—stressed that the articulation of potential damage should be separate from judgments about its significance. The report noted that while complete separation was not possible, a regulative ideal along these lines was

useful (as the subtitle suggests), to guide and manage the process. Scientists should try to focus on the scientific aspects of risk, rather than the cultural, ethical, legal or financial ones, at least in principle.

50. Acid Precipitation Act of 1980, Public Law 96-294, *US Statutes at Large* 94 (1980), 771."

51. US NAS, Committee on Impacts of Stratospheric Change, *Halocarbons*, vii–viii.

52. Robert T. Watson, interview by Keynyn Brysse, Norwich, UK, March 13, 2009; full quotation in chapter 3. Perhaps Watson is correct that governments prefer an implicit approach. Certainly it can be viewed as more respectful of governmental authority and prerogatives than the alternatives. In this sense, the implicit rather than explicit approach may be a useful and defensible rhetorical strategy. It may also serve as a corrective to the common accusation that scientists are arrogant. But if the expectation is that this demarcation will protect the integrity of scientific work and the reputations of the scientists who are doing it, that expectation is likely to be disappointed wherever the policy implications of the findings are clear. The implications of the CIAP report were certainly clear to those who supported the SST program; that is why they objected to it.

53. Jonathan Shanklin and Michael McIntyre, seminar/group interview by Keynyn Brysse, Cambridge UK, March 16, 2009; quoted more extensively in chapter 3.

54. Keller, "What Are Climate Scientists to Do?"

55. Oreskes and Conway, *Merchants of Doubt*, chapter 4.

56. Parson, *Protecting the Ozone Layer*, 41.

57. The issue of neutrality arises in sponsored research as well. In 1976, a group of oceanographers at the Woods Hole Oceanographic Institution became involved in a program to evaluate deep-sea disposal of radioactive waste, sponsored by the Department of Energy through Sandia National Laboratory. One scientist, Vaughan Bowen, likened advocacy to salesmanship and argued they would lose their credibility if they allowed themselves to slide into that mode or even to claim that they were demonstrating anything. "I can't emphasize—evidently—too loudly or too often that we are not involved, and must not represent ourselves as being involved in *demonstrating* (or in selling) *anything*! What we are trying to do is to evaluate the suitability and feasibility of the use of the deep ocean for waste emplacement. If we let 'selling' terminology creep into [our work] then our scientific credibility disappears just as if we had peddled bottles of snake oil!" Vaughan Bowen to D. Richard Anderson, February 6, 1976, Woods Hole Oceanographic Institution Biographical Files, Vaughan Bowen (italics in original). He contrasted evaluation with demonstration, the former being preferable, as it suggested that they were open-minded, the latter implying a formed conclusion. The project under discussion was, however, very much a matter of salesmanship: promoting the idea of deep-sea waste disposal in order to justify the research relevant to it. See Oreskes, *Science on a Mission*, chapter 8.

58. Z. Wang, "Responding to Silent Spring."

59. Robock, "Nuclear Winter."

60. It is often suggested that Sagan's public pronouncements caused him to be denied membership in the US National Academy of Sciences. See, for example, Annalee Newitz, "Why Was Carl Sagan Blackballed from the National Academy of Sciences?" *Gizmodo*, April 27, 2015, http://gizmodo.com/why-was-carl-sagan-blackballed-from-the-national-academ-1700524296 (accessed January 16, 2017). Oreskes and Conway suggest that a more important factor was the role of Fred Seitz in blocking his nomination over the issue of climate change. See Oreskes and Conway, *Merchants of Doubt*, chapter 6.

61. In off-the-record conversations, several scientists have claimed that climate modeler James Hansen has been "discredited" because of his political activism, but this is a claim about how he is viewed by his colleagues, not evidence of how he is viewed by political leaders or others involved in policy formation, much less of what impact his participation would have in an assessment.

62. See IPCC, *Climate Change 1995*, section 4, p. 22. For Santer's defense of himself, see "Ben Santer's Open Letter to the Climate Science Community," http://www.odlt.org/dcd/docs/Ben%20

Santer%20-%20Open%20letter%20to%20the%20climate%20science%20community.pdf (accessed January 16, 2017). Santer recounted these events at the 2011 Stephen H. Schneider Symposium—see http://www2.ucar.edu/for-staff/daily/announcements/2011-06-21/2011-stephen-h-schneider-symposium (accessed January 16, 2017)—but his remarks do not appear to be publicly available.

63. On motivations of those who attack science, see Oreskes and Conway, *Merchants of Doubt*; Michaels, *Doubt Is Their Product*; McCright and Dunlap, "Challenging Global Warming as a Social Problem"; McCright and Dunlap, "Cool Dudes." On the role of ideologically oriented think tanks and their funders, see Medvetz, *Think Tanks in America*; and Brulle, "Institutionalizing Delay."

64. M. Mann, *The Hockey Stick and the Climate Wars.*

65. John M. Broder, "Scientists Taking Steps to Defend Work on Climate," *New York Times*, March 2, 2010, http://www.nytimes.com/2010/03/03/science/earth/03climate.html?_r=0 (accessed January 16, 2017).

66. Mann's emerging activism on climate change followed the attack on his work. See M. Mann, *The Hockey Stick and the Climate Wars.*

67. Oreskes and Conway, *Merchants of Doubt*; Medvetz, *Think Tanks in America*; Brulle, "Institutionalizing Delay."

68. Markowitz and Rosner, *Deceit and Denial*; Markowitz and Rosner, *Lead Wars*; Michaels, *Doubt Is Their Product*; Mnookin, *Panic Virus.*

69. This is prevailing practice at the US National Academy of Science in the selection of both panel participants and reviewers. In forming an NRC panel, staffers will make an effort to "seek balance" in the known views of participants, and when a draft report is ready for review, the committee chair or staff will seek reviewers who they either know or suspect will disagree with the report's conclusion. Participants generally see this as an effective means to improve the quality of the report and to ensure that diverse perspectives are adequately considered and potential arguments and refutations addressed. Matthew Shindell, study of NAS-NRC assessments, in progress; Chris Whipple, written comments on an early draft of this chapter, May 30, 2016.

70. This vision of objectivity is founded on the work of such philosophers as J. S. Mill, C. S. Peirce, and Sir Karl Popper. For an excellent discussion, see Longino, *Science as Social Knowledge.*

71. Although it was rejected by philosophers of science who focused on the question of scientific method as the basis for the reliability of scientific knowledge. See Zamitto, *A Nice Derangement of Epistemes*; Oreskes, *Should We Trust Science?*

72. Oreskes and Conway, *Merchants of Doubt.*

73. Pearce, *Climate Files.*

74. Today most climate scientists would say something similar about anthropogenic climate change: that if greenhouse gas emissions continue to increase, anthropogenic climate change will accelerate and large-scale, harmful impacts will occur, tacitly implying without explicitly stating that greenhouse gas emissions need to be curtailed.

75. As noted in the introduction to this chapter, the scientific effort to distinguish between science and policy closely mirrors the traditional distinction between facts and values. While there has been an enormous amount of ink spilled over the fact/value distinction, including on whether it exists, most scientists have no doubt that it does and is important. For further discussion, see Jamieson, *Reason in a Dark Time*, chapter 3, section 3.3.

76. For the distinction between the total intellectual and the specific expert see Pierre Bourdieu, *The Logic of Practice*, 1–2. For the domain-general/domain-specific distinction, see Hirschfeld and Gelman, *Mapping the Mind*. An overview can be found under the subheading "The Massive Modularity Hypothesis," in "Evolutionary Psychology," *Stanford Encyclopedia of Philosophy*, revised May 21, 2014, https://plato.stanford.edu/entries/evolutionary-psychology/#MasModHyp (accessed January 16, 2017).

77. Quoted in Andrew C. Revkin, "Scientist at Work: Susan Solomon; Melding Science and Diplomacy to Run a Global Climate Review," *New York Times*, February 6, 2007, http://query.nytimes.com/gst/fullpage.html?res=9C06E3DA133FF935A35751C0A9619C8B63 (accessed January 16, 2017). Bob Watson criticized Solomon for "ducking the question of what is needed" while at the same time acknowledging that she "could argue that her neutrality on the policy question provides her greater credibility as an unbiased scientist and [IPCC working group co-chair]." Solomon's comments generated considerable discussion, both on her specific position and on the general issue of scientists as advocates; see, for example, Andrew C. Revkin, "The Road from Climate Science to Climate Activism," *New York Times*, January 9, 2008, http://dotearth.blogs.nytimes.com/2008/01/09/the-road-from-climate-science-to-climate-advocacy/ (accessed January 16, 2017); Andrew C. Revkin, "From Climate Science to Climate Activism—The Sequel," *New York Times*, August 25, 2010, http://dotearth.blogs.nytimes.com/2010/08/25/from-climate-science-to-climate-advocacy-the-sequel/ (accessed January 16, 2017); Rick Piltz, "An Eminent Climate Scientist Working to Hold Government Officials Accountable," Climate Science & Policy Watch, February 12, 2008, http://www.climatesciencewatch.org/2008/02/12/an-eminent-climate-scientist-working-to-hold-government-officials-accountable/ (accessed January 16, 2017); Michael E. Mann, "If You See Something, Say Something," New York Times, January 17, 2014, http://www.nytimes.com/2014/01/19/opinion/sunday/if-you-see-something-say-something.html (accessed January 16, 2017); Eli Kintisch, "Hansen's Climate Science and Advocacy Project Under Way," Science magazine blog, February 26, 2014, http://www.sciencemag.org/news/2014/02/hansens-climate-science-and-advocacy-project-under-way (accessed January 16, 2017); David Ropeik, "Scientists as Scientists, or Advocates, or Both?," Big Think, n.d., http://bigthink.com/risk-reason-and-reality/scientists-as-scientists-or-advocates-or-both (accessed January 16, 2017); and John Schwartz, "Katharine Hayhoe, a Climate Explainer Who Stays above the Storm," New York Times, October 10, 2016, http://www.nytimes.com/2016/10/11/science/katharine-hayhoe-climate-change-science.html (accessed January 16, 2017). James Hansen has criticized the climate science community in general for its "reticence" (Hansen, "Scientific Reticence and Sea Level Rise.")

78. Bourdieu, *The Logic of Practice*, 2. The overconfidence effect is one of the most robust results in social psychology. For a popular discussion, see Kahneman, *Thinking, Fast and Slow*. A thrust of the argument by Oreskes and Conway, *Merchants of Doubt*, is to suggest that we should be troubled when scientists speak assertively on questions outside their specific expertise, as when a physicist makes claims about tobacco control or a climate modeler recommends nuclear energy policy. In the history of nuclear power, there is a long history of physicists overestimating the promise and underestimating the social and economic obstacles. See also Naomi Oreskes, "We Need a New Manhattan Project to Deal with Climate Change," *New York Times*, November 14, 2013, http://www.nytimes.com/roomfordebate/2013/11/14/is-nuclear-power-the-answer-to-climate-change/we-need-a-new-manhattan-project-to-deal-with-climate-change (accessed January 16, 2017). Scientists who thought nuclear power would be too cheap to meter were focused almost exclusively on the huge amounts of energy released by nuclear fission and paid little attention to the engineering challenges of doing so safely, the social challenges of overcoming the shadow of the mushroom cloud, and the downstream problem of radioactive waste disposal.

79. Marvin Goldberger, personal communication, 2007.

80. Economics is sometimes included, the notable exception among the social sciences; why this is so invites further research.

81. IPCC, "Summary for Policymakers." On sustainable development as a social and economic concept, see the UN sustainable development goals, available at https://www.un.org/sustainabledevelopment/sustainable-development-goals/ (accessed April 16, 2018). See also Jamieson, "Sustainability and Beyond."

82. Brandt, *Cigarette Century*; Proctor, *Golden Holocaust*. There is no good scholarly history of asbestos as yet, but useful information can be found in Castelman, *Asbestos*.

83. Proctor, *Value-Free Science?*; Jasanoff, *Designs on Nature*; Douglas, *Science, Policy, and the Value-Free Ideal.*

84. See Meyer et al., "Above the Din but in the Fray."

CHAPTER SIX

1. Intergovernmental Panel on Climate Change (IPCC), "Organization," http://ipcc.ch/organization/organization.shtml (accessed January 16, 2017).

2. The IPCC established the position of "review editor," a senior expert to oversee chapter authors' responses to reviewer comments, to address this concern. While this is a step toward creating an independent arbiter, the IPCC authors nevertheless retain access to the arbiter and leverage over the outcome that far exceeds that of authors of journal submissions.

3. Jamieson, "Scientific Uncertainty and the Political Process."

4. National Research Council (NRC), Committee on Atmospheric Transport and Chemical Transformation in Acid Precipitation, *Acid Deposition.*

5. NASA et al., *Scientific Assessment of Stratospheric Ozone*; NASA et al., *Scientific Assessment of Ozone Depletion: 1991.*

6. NASA/WMO, *Atmospheric Ozone 1985.*

7. Brysse et al., "Climate Change Prediction."

8. IPCC, *IPCC Special Report on Emissions Scenarios.*

9. Rahmstorf, "Semi-Empirical Approach to Projecting Future Sea-Level Rise"; Vaughan and Spouge, "Risk Estimation of Collapse of the West Antarctic Ice Sheet"; Oppenheimer et al., "Climate Change."

10. Rahmstorf, "Semi-Empirical Approach to Projecting Future Sea-Level Rise."

11. Gregory et al., "Twentieth-Century Global-Mean Sea Level Rise."

12. Oppenheimer, O'Neill, and Webster, "Negative Learning."

13. Kennel and Daultrey, *Knowledge Action Networks.*

14. The error was the assertion in the *Fourth Assessment Report* that the area of the Himalayas covered by glaciers would shrink by 80% by 2035. See Kargel, "Himalayan Glaciers."

15. Richard Black, "UN Climate Body Admits 'Mistake' on Himalayan Glaciers," *BBC News*, http://news.bbc.co.uk/2/hi/science/nature/8468358.stm (accessed January 21, 2017); Kargel, "Himalayan Glaciers."

CONCLUSION

1. On the question of who is an expert and what constitutes expertise, see Collins and Evans, *Rethinking Expertise*. For general background on the relation between scientists, governance, and democracy, see Kitcher, *Science, Truth, and Democracy*; Jasanoff, *Designs on Nature*; Jasanoff, *Science and Public Reason*; Lentsch and Weingart, *The Politics of Scientific Advice.*

2. For more on the IPCC, see Vardy et al., "The Intergovernmental Panel on Climate Change."

3. On the production of ignorance, see Proctor and Schiebinger, *Agnotology.*

4. Richardson and Wæver, in "Building Bridges between Scientists and Policymakers," argue that a key reason assessments have emerged as important activities in recent decades may be that the gap between the needs of decision makers and what academic science can provide has grown.

5. For an alternative view, see Kowarsch, *A Pragmatist Orientation for the Social Sciences in Climate Policy.*

6. On methodological preference as a source of epistemic bias, see Oreskes, *The Rejection of Continental Drift*; and Oreskes, *Should We Trust Science?*

7. Mike Hulme and Martin Mahoney, in "Climate Change," note that social science literature is conspicuously lacking from most of the citations in IPCC reports—a result that might be predictable given the lack of social scientists on the panels. Drawing on the work of Bjurstrom and Polk, "Physical and Economic Bias in Climate Change Research," they note that of the fourteen thousand references cited in the IPCC *Third Assessment Report*, only 12% of the peer-reviewed papers cited were from the social sciences; if economics is excluded, that figure drops to less than 8%. Bias in the choice of experts results in at least potential bias, or blind spots, in the use and understanding of pertinent literature. Others who have noted a "natural science" bias in the articulation and analysis of the "climate change problem"—and the ways in which social considerations are constructed as secondary in IPCC analyses—include Malone and Rayner, "The Role of the Research Standpoint in Integrating Global-Scale and Local-Scale Research"; Miller, "Climate Science and the Making of a Global Political Order,"; Shackley and Skodvin, "IPCC Gazing and the Interpretive Social Sciences"; and Yearley, "Sociology and Climate Change after Kyoto."

8. In a review of how experts were recruited for the Millennium Ecosystem Assessment, Leifeld and Fisher, in "Membership Nominations in International Scientific Assessments," found that personal affinity, including prior service on an assessment or in an elite international network, institution, or organization, was the strongest predictor of inclusion; people tend to nominate people they already know. In particular, individuals have a great tendency to nominate other individuals of the same nationality or who have the same employer or university affiliation. They suggest that this may have negative "repercussions." Our observations are consistent with their conclusion that personal connections and prior acquaintance play a significant part in who is chosen to participate in an assessment. On the other hand, given the nature of expert communities, it seems almost inevitable that most relevant experts will already know each other or have some prior personal or institutional connection. It is difficult to imagine how one could find experts who do not have at least some degree of personal affinity with each other.

9. See, for example, Bloor, *Knowledge and Social Imagery*; and Vaughan, *The Challenger Launch Decision*.

10. Wormbs and Sörlin, "Arctic Futures."

11. Parson, *Protecting the Ozone Layer*.

12. The goal of being "policy-relevant but not policy-prescriptive" was used in other assessments as well as the IPCC's—notably, in the international assessments regarding ozone depletion—but in the IPCC specifically this maxim has been formalized as an official guiding principle. See IPCC, "Statement on IPCC Principles and Procedures," February 2, 2010, http://www.ipcc.ch/pdf/press/ipcc-statement-principles-procedures-02-2010.pdf (accessed January 28, 2017).

13. Gieryn, "Boundary-Work and the Demarcation of Science from Non-Science." See also Gieryn, "Boundaries of Science."

14. For one thoughtful perspective on the complex entanglements of science, politics, and policy, see Weingart, "The Moment of Truth for Science."

15. Oreskes, "The Scientist as Sentinel"; Oreskes, Jamieson, and Oppenheimer, "What Role for Scientists?" See also Naomi Oreskes, "The Scientist as Sentinel," Harvard University Faculty of Arts and Sciences, Division of Science, Science Research Public Lecture, March 29, 2017, https://www.youtube.com/watch?v=H5QQ5UyjVgA (accessed January 2, 2018).

16. On the problem of reticence, see Hansen, "Scientific Reticence and Sea Level Rise." For a recent paper calling for scientific assessments to be "solution" oriented and to embrace policy, see Kowarsch et al., "A Road Map for Global Environmental Assessments."

17. On the performative aspects of assessments, see Hilgartner, *Science on Stage*; and Hulme, "The Performance of Science."

18. Compare this to the old saw attributed to Daniel Patrick Moynihan that "everyone is entitled to their own opinion, but not to their own facts."

19. Latour, "Why Has Critique Run Out of Steam?"

20. Brysse et al., "Climate Change Prediction."

21. Cash et al., "How Can Assessment Processes and Outcomes Be Improved?"

22. Patt, *Assessing Extreme Outcomes.*

23. Urofsky, *Dissent and the Supreme Court.*

24. For an early critique of such idealized models in the context of climate change, see Jamieson, "Managing the Future."

25. There is a substantial literature on this point. An entry point is Hulme and Mahoney, "Climate Change," in which they note that previous studies show that 80%–82% of IPCC authors come from Organisation for Economic Co-operation and Development countries and suggest that this has negative implications for how well IPCC findings are trusted, particularly by those who feel excluded by the process. See also Lahsen, "Transnational Locals"; and Lahsen, "Trust through Participation?" Available evidence suggests that, despite efforts to create diverse panels, Anglophone scientists from Europe and North America still predominate in the IPCC. See Ho-Lem, Zerriffi, and Kandlikar, "Who Participates in the Intergovernmental Panel on Climate Change and Why"; and Hulme, "The Performance of Science."

26. On the impact of the IPCC on scientific knowledge production outside the assessment itself, see Vasileiadou, Heimeriks, and Petersen, "Exploring the Impact of the IPCC Assessment Reports on Science." These authors investigate how often IPCC reports are cited, find that the answer is very often, and conclude that the IPCC is an "epistemological monster" that has had "a considerable impact to [*sic*] climate change science" (1056) and that this impact "can be understood as the utilization of knowledge claims of the IPCC reports in subsequent scientific results" (1059).We agree with these authors that the IPCC is a "contributor of knowledge claims," not merely a compiler of them. However, we would suggest that the high citation rate only tells us that scientists find it useful and appropriate to cite IPCC reports; it does not tell us in what way those reports are being used.

27. On the production of ignorance via undue focus on areas of immediate geopolitical concern during the Cold War, see Oreskes, *Science on a Mission.*

Bibliography

AAAS (American Association for the Advancement of Science). 1980. *Conference Proceedings*. The Conference of the American Association for the Advancement of Sciencc Studying the Response of the West Antarctic Ice Sheet to CO_2-Induced Climatic Warming. April 8–10, 1980, University of Maine at Orono.

Aaserud, Finn. 1999. "The Scientist and the Statesmen: Niels Bohr's Political Crusade during World War II." *Historical Studies in the Physical and Biological Sciences* 30(1): 1–47.

Ad Hoc Committee on Acid Rain: Science and Policy. 1985. *Is There Scientific Consensus on Acid Rain? Excerpts from Six Governmental Reports*, prepared by Charles T. Driscoll, James N. Galloway, James F. Hornig, Gene E. Likens, Michael Oppenheimer, Kenneth A. Rahn, and David W. Schindler. Millbrook, NY: Institute of Ecosystem Studies.

Albritton, Daniel. 1998. "What Should Be Done in a Science Assessment?" In *Protecting the Ozone Layer: Lessons, Models, and Prospects*, edited by Phillipe G. LePrestre, John D. Reid, and E. Thomas Morehouse Jr., 67–74. Boston: Kluwer Academic.

Alder, Ken. 2010. *Engineering the Revolution: Arms and Enlightenment in France, 1763–1815*. Chicago: University of Chicago Press.

Alperovitz, Gar. 1995. *The Decision to Use the Atomic Bomb*. New York: Random House.

Andersen, Stephen O., K. Madhava Sarma, and Lani Sinclair. 2002. *Protecting the Ozone Layer: The United Nations History*. London: Earthscan.

Aronova, Elena, Karen S. Baker, and Naomi Oreskes. 2010. "Big Science and Big Data in Biology: From the International Geophysical Year through the International Biological Program to the Long Term Ecological Research (LTER) Network, 1957–Present." *Historical Studies in the Natural Sciences* 40(2): 183–224.

Balogh, Brian. 1991. *Chain Reaction: Expert Debate and Public Participation in American Commercial Nuclear Power, 1945–1975*. Cambridge: Cambridge University Press.

Baulch, D. L., R. A. Cox, R. F. Hampson Jr., J. A. Kerr, J. Troe, and R. T. Watson. 1980. "Evaluated Kinetic and Photochemical Data for Atmospheric Chemistry." *Journal of Physical and Chemical Reference Data* 9(2): 295–472.

Benedick, Richard Elliot. 1991. *Ozone Diplomacy: New Directions in Safeguarding the Planet*. Cambridge, MA: Harvard University Press.

Bentley, C. R. 1982. "The West Antarctic Ice Sheet: Diagnosis and Prognosis." In *Proceedings: Carbon Dioxide Research Conference: Carbon Dioxide, Science and Consensus, September 19–23, 1982, Berkeley Springs, West Virginia*, IV.3–IV.50. Washington, DC: Department of Energy.

Berg, Paul, and Maxine F. Singer. 1995. "The Recombinant DNA Controversy: Twenty Years Later." *Proceedings of the National Academy of Sciences USA* 92: 9011–9013.

Bernstein, Barton, and Peter Galison. 1989. "In Any Light: Scientists and the Decision to Build the Superbomb, 1952–1954." *Historical Studies in the Physical Sciences* 19(2): 267–347.

Betsill, M., and R. Pielke Jr. 1998. "Blurring the Boundaries: Domestic and International Ozone Politics and Lessons for Climate Change." *International Environmental Affairs* 10: 147–172.

Bird, Kai, and Martin Sherwin. 2005. *American Prometheus: The Triumph and Tragedy of J. Robert Oppenheimer*. New York: Vintage.

Bjurstrom, Andreas, and Merritt Polk. 2011. "Physical and Economic Bias in Climate Change Research: A Scientometric Study of IPCC Third Assessment Report." *Climatic Change* 108(1–2): 1–22.

Bloor, David. (1976) 1991. *Knowledge and Social Imagery*. 2nd ed. Chicago: University of Chicago Press.

Bolin, B. 2007. *A History of the Science and Politics of Climate Change: The Role of the Intergovernmental Panel on Climate Change*. Cambridge: Cambridge University Press.

Bourdieu, Pierre. 1990. *The Logic of Practice*. Stanford, CA: Stanford University Press.

Brandt, Allan M. 2007. *The Cigarette Century: The Rise, Fall, and Deadly Persistence of the Product That Defined America*. New York: Basic Books.

Brimblecombe, Peter. 1987. *The Big Smoke: A History of Air Pollution in London since Medieval Times*. London: Methuen.

Brulle, Robert J. 2014. "Institutionalizing Delay: Foundation Funding and the Creation of US Climate Change Counter-Movement Organizations." *Climatic Change* 122(4): 681–694.

Brysse, Keynyn, Naomi Oreskes, Jessica O'Reilly, and Michael Oppenheimer. 2013. "Climate Change Prediction: Erring on the Side of Least Drama?" *Global Environmental Change* 23(1): 327–337.

Budd, W. F., D. Jenssen, and I. N. Smith. 1984. "A Three-Dimensional Time-Dependent Model of the West Antarctic Ice Sheet." *Annals of Glaciology* 5: 29–36.

Burke, John G. 1966. "Bursting Boilers and the Federal Power." *Technology and Culture* 7(1): 1–23.

Cabantous, L. 2007. "Ambiguity Aversion in the Field of Insurance: Insurers' Attitudes to Imprecise and Conflicting Probability Estimates." *Theory and Decision* 62(3): 219–240.

Caciola, Nancy. 2015. *Discerning Spirits: Divine and Demonic Possession in the Middle Ages*. Ithaca, NY: Cornell University Press.

Cadle, R. D., P. Crutzen, and D. Ehhalt. 1975. "Heterogeneous Chemical Reactions in the Stratosphere." *Journal of Geophysical Research* 24: 3381–3385.

Callis, L. B., and M. Natarajan. 1986. "The Antarctic Ozone Minimum: Relationship to Odd Nitrogen, Odd Chlorine, the Final Warming, and the 11-Year Solar Cycle." *Journal of Geophysical Research* 91: 10771–10796.

Canada/United States Coordinating Committee. 1982/1983. *United States–Canada Memorandum of Intent on Transboundary Air Pollution: Final Report*. 3 vols. Downsview, ON: Environment Canada/Washington, DC: US Environmental Protection Agency.

Carson, Rachel. 1962. *Silent Spring*. Boston: Houghton Mifflin.

Cash, David W., and William C. Clark. 2001. *From Science to Policy: Assessing the Assessment Process*. John F. Kennedy School of Government, Harvard University, Faculty Research Working Paper 01-045. Cambridge, MA: Kennedy School of Government, Harvard University. http://www.hks.harvard.edu/gea/pubs/rwp01-045.htm (accessed January 28, 2017).

Cash, David W., K. Fisher-Vanden, W. Franz, R. Frosch, J. Holdren, J. Jaeger, M. Kandlikar, A. Sagar, and R. Stavins. 1997. "How Can Assessment Processes and Outcomes Be Improved?" In *Global Environmental Assessment Project, 1997: A Critical Evaluation of Global Environmental Assessments, the Climate Experience*, edited by William C. Clark, James McCarthy, and Eileen Shea, 117–143. Calverton, MD: CARE.

Castelman, Barry I. 2004. *Asbestos: Medical and Legal Aspects*. Aspen, CO: Aspen Publishers.

CDAC (Carbon Dioxide Assessment Committee), NRC (National Research Council). 1982. *Increasing Carbon Dioxide and the West Antarctic Ice Sheet: Notes on an Informal Workshop*. San Diego, CA: Scripps Institution of Oceanography.

———. 1983. *Changing Climate*. Washington, DC: National Academy Press.

Chen, J. L., C. R. Wilson, D. Blankenship, and B. D. Tapley. 2009. "Accelerated Antarctic Ice Loss from Satellite Gravity Measurements." *Nature Geoscience* 2: 859–862.

Christie, Maureen. 2001. *The Ozone Layer: A Philosophy of Science Perspective*. Cambridge: Cambridge University Press.

Churchill, Randolph S. 1964. *Twenty-One Years*. London: Weidenfeld and Nicolson.

Collins, Harry, and Robert Evans. 2007. *Rethinking Expertise*. Chicago: University of Chicago Press.

Committee on Sulfur Oxides, Board on Toxicology and Environmental Health Hazards, Assembly of Life Sciences, NRC. 1978. *Sulfur Oxides*. Washington, DC: National Academy of Sciences.

Commoner, Barry. 1974. *The Closing Circle: Nature, Man and Technology*. New York: Bantam Books.

Conway, Erik M. 2005. *High-Speed Dreams: NASA and the Technopolitics of Supersonic Transportation, 1945–1999*. Baltimore: Johns Hopkins University Press.

Cowling, Ellis B. 1978. "Effects of Acid Precipitation and Atmospheric Deposition on Terrestrial Vegetation." In *A National Program for Assessing the Problem of Atmospheric Deposition (Acid Rain): A Report to the Council on Environmental Quality*, edited by James Galloway, Ellis B. Cowling, Eville Gorham, and William McFee, 46–63. National Atmospheric Deposition Program.

———. 1992. "The Performance and Legacy of NAPAP." *Ecological Applications* 2(2): 111–116.

Cowling, Ellis B., and Jan Nilsson. 1992. "A Comparison of Some National Assessments." In *Acidification Research: Evaluation and Policy Applications*, edited by T. Schneider, 279–92. Amsterdam: Elsevier.

Crosland, Maurice. 2002. *Science under Control: The French Academy of Sciences, 1795–1914*. New York: Cambridge University Press.

Crutzen, P. J. 1970. "The Influence of Nitrogen Oxides on the Atmospheric Ozone Content." *Quarterly Journal of the Royal Meteorological Society* 96: 320–325.

Daniel, J. S., S. Solomon, and D. L. Albritton. 1995. "On the Evaluation of Halocarbon Radiative Forcing and Global Warming Potentials." *Journal of Geophysical Research* 100: 1271–1285.

DeAngelis, Hernan, and Pedro Skvarca. 2003. "Glacial Surge after Ice Sheet Collapse." *Science* 299: 1560–1562.

Dochinger, Leon S., and Selinga, Thomas A., eds. 1976. *Proceedings of the First International Symposium on Acid Precipitation and the Forest Ecosystem*. Upper Darby, PA: Forest Service, US Dept. of Agriculture, Northeastern Forest Experiment Station.

Doel, Ronald E. 2003. "Constituting the Postwar Earth Sciences: The Military's Influence on the Environmental Sciences in the USA after 1945." *Social Studies of Science* 33(5): 635–666.

Douglas, Heather E. 2009. *Science, Policy, and the Value-Free Ideal*. Pittsburgh: University of Pittsburgh Press.

Dupree, A. Hunter. 1957. "The Founding of the National Academy of Sciences: A Reinterpretation." *Proceedings of the American Philosophical Society* 101(5): 434–440.

EDF (Environmental Defense Fund). 1988. Testimony of Joseph Goffman, Staff Attorney, Environmental Defense Fund, Before the Subcommittee on Natural Resources, Agricultural Research and Environment, Committee on Science, Space and Technology, House of Representatives, April 27, 1988, US Congress (House), 130–136.

Edwards, Paul N. 1996. *The Closed World: Computers and the Politics of Discourse in Cold War America*. Cambridge, MA: MIT Press.

Ehrlich, Paul. 1970. *The Population Bomb*. New York: Sierra Club / Ballantine Books.

Farley, John, and Gerald L. Geison. 1974. "Science, Politics and Spontaneous Generation in Nineteenth-Century France: The Pasteur-Pouchet Debate." *Bulletin of the History of Medicine* 48(2): 161–198.

Farman, J. C., B. G. Gardiner, and J. D. Shanklin. 1985. "Large Losses of Total Ozone in Antarctica Reveal Seasonal ClO_x/NO_x Interaction." *Nature* 315: 207–210.

Fastook, J. L. 1984. "West Antarctica, the Sea-Level Controlled Marine Instability: Past and Future." In *Climate Processes and Climate Sensitivity*, edited by James E. Hansen and Taro Takahashi, 275–287. American Geophysical Union.

Finkbeiner, Ann. 2007. *The Jasons: The Secret History of Science's Postwar Elite*. New York: Penguin.

Fleming, James. 1998. *Historical Perspectives on Climate Change*. Oxford: Oxford University Press.

Forman, Paul. 1987. "Behind Quantum Electronics: National Security as Basis for Physical Research in the United States, 1940–1960." *Historical Studies in the Physical and Biological Sciences* 18(1): 149–229.

Forster, Bruce A. 1993. *The Acid Rain Debate: Science and Special Interests in Policy Formation*. Ames: Iowa State University Press.

Frankland, Percy. 1898. *Pasteur*. London: Cassell.

Galison, Peter. 1997. *Image and Logic: A Material Culture of Microphysics*. Chicago: University of Chicago Press.

Galison, Peter, and Bruce William Hevly. 1992. *Big Science: The Growth of Large-Scale Research*. Stanford, CA: Stanford University Press.

Galloway, James N. 1988. Testimony of James N. Galloway, Professor, Environmental Sciences Department, University of Virginia, to the Subcommittee on Natural Resources, Agricultural Research, and Environment, Committee on Science, Space and Technology, US House of Representatives, Washington, DC, Wednesday, April 27, 1988. US Congress (House), 119–127.

Galloway, James N., Ellis B. Cowling, Eville Gorham, and William W. McFee. 1978. *A National Program for Assessing the Problem of Atmospheric Deposition (Acid Rain): A Report to the Council on Environmental Quality*. Fort Collins, CO: National Atmospheric Deposition Program.

Gieryn, Thomas F. 1983. "Boundary-Work and the Demarcation of Science from Non-Science: Strains and Interests in Professional Ideologies of Scientists." *American Sociological Review* 48(6): 781–795.

———. 1997. "Boundaries of Science." In *Science and the Quest for Reality*, edited by Alfred I. Tauber, 293–332. London: Macmillan.

Glantz, M., J. Robinson, and M. Krenz. 1982. "Improving the Science of Climate-Related Impact Studies: A Review of Past Experience." In *Carbon Dioxide Review: 1982*, edited by William Clark, 58–93. New York: Oxford University Press.

Goldberg, Stanley. 1992. "Creating a Climate of Opinion: Vannevar Bush and the Decision to Build the Bomb." *Isis* 83: 429–452.

Gordon, A. L. 1983. "Comments about the Ocean Role in the Antarctic Glacial Balance." In *Proceedings: Carbon Dioxide Research Conference: Carbon Dioxide, Science and Consensus, September 19–23, 1982, Berkeley Springs, West Virginia*, IV.75–IV.85. Washington, DC: Department of Energy.

Gornitz, V., S. Lebedeff, and J. Hansen. 1982. "Global Sea Level Trend in the Past Century." *Science* 215: 1611–1614.

Gregory, J. M., et al. 2013. "Twentieth-Century Global-Mean Sea Level Rise: Is the Whole Greater Than the Sum of the Parts?" *Journal of Climate* 26: 4476–4499.

Grundmann, Reiner. 2001. *Transnational Environmental Policy: Reconstructing Ozone*. Routledge Studies in Science, Technology, and Society. London: Routledge.

Hales, Jeremy. 1976. *Tall Stacks and the Atmospheric Environment*. Research Triangle Park, NC: Environmental Protection Agency.

Halvorsen, T. 1987. "An Alarming and Mysterious 'Hole' in Antarctica's Protective Ozone Shield Is Getting Worse and May Have Global Ramifications, a Group of Scientists Reported Wednesday." *USA Today*, September 30.

Hamblin, Jacob Darwin. 2013. *Arming Mother Nature: The Birth of Catastrophic Environmentalism*. New York: Oxford University Press.

Hammonds, Evelynn M. 2002. *Childhood's Deadly Scourge: The Campaign to Control Diphtheria in New York City, 1880–1930*. Baltimore: Johns Hopkins University Press.

Hampson, J. 1965. "Chemiluminescent Emission Observed in the Stratosphere and Mesosphere." In *Les Problèmes météorologiques de la stratosphère et de la mésosphère*, 393–440. Paris: Presses Universitaires de France.

Hansen, J. E. 2007. "Scientific Reticence and Sea Level Rise." *Environmental Research Letters* 2(2): 024002.

Harrington, Anne. 1989. *Medicine, Mind, and the Double Brain: A Study in Nineteenth-Century Thought*. Princeton, NJ: Princeton University Press.

Harvey, Richard J., and Harold H. Beamish. 1972. "Acidification of the La Cloche Mountain Lakes, Ontario, and Resulting Fish Mortalities." *Journal of the Fisheries Research Board of Canada* 29: 1131–1143.

Herrick, Charles. 2000. "Predictive Modeling of Acid Rain: Obstacles to Generating Useful Information." In *Prediction: Science, Decision Making, and the Future of Nature*, edited by Daniel Sarewitz, Roger A. Pielke Jr., and Radford Byerly Jr., 251–268. Washington, DC: Island.

Herrick, Charles, and Dale Jamieson. 1995. "The Social Construction of Acid Rain: Some Implications for Science/Policy Assessment." *Global Environmental Change* 5(2): 105–112.

Hewlett, Richard G., and Francis Duncan. 1970. *Atomic Shield: 1947–1952*. Vol. 2 of *A History of the US Atomic Energy Commission*. Washington, DC: US Department of Energy.

Hilgartner, Stephen. 2000. *Science on Stage: Expert Advice as Public Drama*. Stanford, CA: Stanford University Press.

Hilgartner, Stephen, Clark Miller, and Rob Hagendijk. 2015. *Science and Democracy: Making Knowledge and Making Power in the Biosciences and Beyond*. New York: Routledge.

Hirschfeld, Lawrence A., and Susan A. Gelman, eds. 1994. *Mapping the Mind: Domain Specificity in Cognition and Culture*. New York: Cambridge University Press.

Hoffman, John S., and Michael J. Gibbs. 1988. *Future Concentrations of Stratospheric Chlorine and Bromine*. EPA 400/1-88/005. Washington, DC: US EPA, Office of Air and Radiation, July.

Ho-Lem, Claudia, Hisham Zerriffi, and Milind Kandlikar. 2011. "Who Participates in the Intergovernmental Panel on Climate Change and Why: A Quantitative Assessment of the National Representation of Authors in the Intergovernmental Panel on Climate Change." *Global Environmental Change* 21(4): 1308–1317.

Hulme, Mike. 2009. "The Performance of Science." In *Why We Disagree about Climate Change: Understanding Controversy, Inaction and Opportunity*, 72–108. Cambridge: Cambridge University Press.

Hulme, Mike, and Martin Mahony. 2010. "Climate Change: What Do We Know about the IPCC?" *Progress in Physical Geography* 34: 705–718.

Hunt, B. G. 1966. "Photochemistry of Ozone in a Moist Atmosphere." *Journal of Geophysical Research* 71: 1385–1398.

Huybrechts, P. 1990. "A 3-D Model for the Antarctic Ice Sheet: A Sensitivity Study on the Glacial-Interglacial Contrast." *Climate Dynamics* 5(2): 79–92.

Huybrechts, P., and J. Oerlemans. 1990. "Response of the Antarctic Ice Sheet to Future Greenhouse Warming." *Climate Dynamics* 5(2): 93–102.

IEEE (Institute of Electrical and Electronics Engineers) Power Engineering Society. 1981. *Carbon Dioxide Proliferation: Will the Icecaps Melt? Paper Presented at the 1981 Summer Meeting, Portland, OR, July 28, 1981*. New York: IEEE Publication Services.

IIASA (International Institute for Applied Systems Analysis). 1981. *Life on a Warmer Earth: Possible Climatic Consequences of a Man-Made Global Warming*. IIASA Executive Report ER-81-003. Laxenburg, Austria: IIASA.

Institute for Energy Analysis. 1983. *Proceedings: Carbon Dioxide Research Conference: Carbon Dioxide, Science and Consensus, September 19–23, 1982, Berkeley Springs, West Virginia*. Washington, DC: Institute for Energy Analysis.

Interagency Task Force on Acid Precipitation. 1981. *National Acid Precipitation Assessment Plan Draft*. Washington, DC: Interagency Task Force on Acid Precipitation.

———. 1984. *Report on the Second Annual Review Meeting of the National Acid Precipitation Assessment Program: Research to Assist Decisionmaking*. Washington, DC: Interagency Task Force on Acid Precipitation.

IPCC (Intergovernmental Panel on Climate Change). 1990. *Climate Change: The IPCC Scientific Assessment*. Cambridge: Cambridge University Press.

———. 1996. *Climate Change 1995: A Report of the Intergovernmental Panel on Climate Change. Second Assessment Report of the Intergovernmental Panel on Climate Change*. IPCC. https://www.ipcc.ch/pdf/climate-changes-1995/ipcc-2nd-assessment/2nd-assessment-en.pdf (accessed January 24, 2017).

———. 2000. *IPCC Special Report on Emissions Scenarios*. A Special Report of IPCC Working Group III. IPCC. http://www.grida.no/publications/other/ipcc_sr/ (accessed January 17, 2017).

———. 2001. *Climate Change 2001: The Scientific Basis. Contribution of Working Group I to the Third Assessment Report of the Intergovernmental Panel on Climate Change*, edited by J. T. Houghton, Y. Ding, D. J. Griggs, M. Noguer, P. J. van der Linden, X. Dai, K. Maskell, and C. A. Johnson. Cambridge: Cambridge University Press.

———. 2007. *Climate Change 2007: The Physical Science Basis. Contribution of Working Group I to the Fourth Assessment Report of the Intergovernmental Panel on Climate Change*, edited by S. Solomon, D. Qin, M. Manning, Z. Chen, M. Marquis, K. B. Averyt, M. Tignor, and H. L. Miller. Cambridge: Cambridge University Press.

———. 2014. "Summary for Policymakers." In *Climate Change 2014: Mitigation of Climate Change. Contribution of Working Group III to the Fifth Assessment Report of the Intergovernmental Panel on Climate Change*, edited by O. Edenhofer, R. Pichs-Madruga, Y. Sokona, E. Farahani, S. Kadner, K. Seyboth, A. Adler, I. Baum, S. Brunner, P. Eickemeier, B. Kriemann, J. Savolainen, S. Schlömer, C. von Stechow, T. Zwickel, and J. C. Minx. Cambridge: Cambridge University Press.https://www.ipcc.ch/pdf/assessment-report/ar5/wg3/ipcc_wg3_ar5_summary-for-policymakers.pdf (accessed January 16, 2017).

Jamieson, Dale. 1990. "Managing the Future: Public Policy, Scientific Uncertainty, and Global Warming." In *Upstream/Downstream: Essays in Environmental Ethics*, edited by D. Scherer, 67–89. Philadelphia: Temple University Press.

———. 1996. "Scientific Uncertainty and the Political Process." *Annals of the American Academy of Political and Social Sciences* 545 (May): 35–43.

———. 1998. "Sustainability and Beyond." *Ecological Economics* 24: 183–192. Reprinted in *Sustainability*, edited by T. Campbell and D. Mollica, 565–574. Farnham, UK: Ashgate, 1990.

———. 2014. *Reason in a Dark Time: Why the Struggle to Stop Climate Change Failed and What It Means for our Future*. Oxford: Oxford University Press.

———. 2015. "Theology and Politics in *Laudato Si'*." *AJIL Unbound* 109: 122–126.

Jasanoff, Sheila. 2005. *Designs on Nature: Science and Democracy in Europe and the United States*. Princeton, NJ: Princeton University Press.

———. 2012. *Science and Public Reason*. London and New York: Routledge.

Jewett, Frank B. 1962. "The Academy—Its Charter, Its Functions and Relations to Government." *Proceedings of the National Academy of Sciences of the United States of America* 48(4): 481–490.

Kahneman, Daniel. 2011. *Thinking, Fast and Slow*. New York: Farrar, Straus and Giroux.

Kargel, Jeffrey S. 2011. "Himalayan Glaciers: The Big Picture Is a Montage." *Proceedings of the National Academy of Sciences USA* 108(36): 14709–14710.

Keller, Evelyn Fox. 2011. "What Are Climate Scientists to Do?" *Spontaneous Generations* 5(1): 19–26.

Kennel, Charles F. 2015. "Global Knowledge Action Network." In *Proceedings of the Joint Workshop on Sustainable Humanity, Sustainable Nature: Our Responsibility*, edited by Partha S. Dasgupta, Veerabhadran Ramanathan, and Marcelo Sanchez Sorondo, 347–369. Vatican City: Pontifical Academy of Sciences.

Kennel, Charles F., and S. Daultrey. 2010. *Knowledge Action Networks: Connecting Regional Climate Change Assessments to Local Action*. San Diego: UCSD Sustainability Solutions Institute. https://escholarship.org/uc/item/8gd6j0k5 (accessed January 17, 2017).

Kevles, Dan. 1990. "Cold War and Hot Physics: Science, Security, and the American State, 1945–56." *Historical Studies in the Physical and Biological Sciences* 20(2): 239–264.

Kincaid, Harold, John Dupré, and Alison Wylie, eds. 2007. *Value-Free Science? Ideals and Illusions*. Oxford: Oxford University Press.

Kitcher, Philip. 2001. *Science, Truth, and Democracy*. Oxford: Oxford University Press.

Kowarsch, Martin. 2016. *A Pragmatist Orientation for the Social Sciences in Climate Policy: How to Make Integrated Economic Assessments Serve Society*. Boston Studies in the Philosophy and History of Science, vol. 323. Springer International.

Kowarsch, Martin, Jason Jabbour, Christian Flachsland, Marcel T. J. Kok, Robert Watson, Peter M. Haas, Jan C. Minx, Joseph Alcamo, Jennifer Garard, Pauline Riousset, László Pintér, Cameron Langford, Yulia Yamineva, Christoph von Stechow, Jessica O'Reilly, and Ottmar Edenhofer. 2017. "A Road Map for Global Environmental Assessments." *Nature Climate Change* 7: 379–382.

Krieger, Nancy, and Anne-Emanuelle Birn. 1998. "A Vision of Social Justice as the Foundation of Public Health: Commemorating 150 Years of the Spirit of 1848." *American Journal of Public Health* 88(11): 1603–1606.

Krige, John. 2006. *American Hegemony and the Postwar Reconstruction of Science in Europe*. Cambridge, MA: MIT Press.

———. 2016. *Sharing Knowledge, Shaping Europe: US Collaboration and Nonproliferation*. Cambridge, MA: MIT Press.

Kuhn, Thomas S. 1962. *The Structure of Scientific Revolutions*. Chicago: University of Chicago Press.

Kulp, J. Laurence. 1990. "Acid Rain." *Cato Review of Business & Government*, Winter, 41–50.

———. 1994. "Government Measures to Reduce Acid Rain Are Unnecessary." In *Water: Opposing Viewpoints*, edited by Carol Wekesser, 116–119. San Diego: Greenhaven.

Lahsen, Myanna. 2004. "Transnational Locals: Brazilian Experiences of the Climate Regime." In *Earthly Politics: Local and Global in Environmental Governance*, edited by Sheila Jasanoff and M. L. Martello, 151–172. Cambridge, MA: MIT Press.

———. 2007. "Trust through Participation? Problems of Knowledge in Climate Decision-Making." In *The Social Construction of Climate Change: Power, Knowledge, Norms, Discourses*, edited by Mary E. Pettenger, 173–196. Aldershot: Ashgate.

Lambright, Henry W. 2005. *NASA and the Environment: The Case of Ozone Depletion*. Monographs in Aerospace History no. 38. NASA SP-2005-4538. http://history.nasa.gov/monograph38.pdf (accessed November 26, 2012).

Latour, Bruno. 1993. *We Have Never Been Modern*. Cambridge, MA: Harvard University Press.

———. 2004. *Politics of Nature: How to Bring the Sciences into Democracy*. Cambridge, MA: Harvard University Press.

———. 2004. "Why Has Critique Run Out of Steam? From Matters of Fact to Matters of Concern." *Critical Inquiry* 30 (Winter): 225–248.

Leifeld, Philip, and Dana R. Fisher. 2017. "Membership Nominations in International Scientific Assessments." *Nature Climate Change* 7: 730–735.

Lentsch, Justus, and Peter Weingart, eds. 2011. *The Politics of Scientific Advice: Institutional Advice for Quality Assurance*. Cambridge: Cambridge University Press.

LePrestre, Phillipe G., John D. Reid, and E. Thomas Morehouse Jr., eds. 1998. *Protecting the Ozone Layer: Lessons, Models, and Prospects*. Boston: Kluwer Academic.

Lewandowsky, Stephan, James S. Risby, and Naomi Oreskes. 2016. "The 'Pause' in Global Warming: Turning a Routine Fluctuation into a Problem for Science." *Bulletin of the American Meteorological Society* 97: 723–733.

Likens, Gene E. 1984. "Acid Rain: The Smokestack Is the 'Smoking Gun.'" *Garden* 8(4): 12–18.

———. 1988. Letter from Gene E. Likens to Congressman John D. Dingell. US Congress (House), 65–71.

———. 1988. Testimony to the Subcommittee on Natural Resources, Agriculture Research and Environment on the National Acid Precipitation Assessment Program (NAPAP), 27 April 1988. US Congress (House), 55–64.

———. 2010. "The Role of Science in Decision Making: Does Evidence-Based Science Drive Environmental Policy?" *Frontiers in Ecology* 8(6): e1–e9.

Likens, Gene E., and Hermann F. Bormann. 1974. "Acid Rain: A Serious Regional Environmental Problem." *Science* 184: 1176–1179.

Likens, Gene E., Hermann F. Bormann, and Noye Johnson. 1972. "Acid Rain." *Environment* 14(2): 33–40.

Lingle, Craig S. 1984. "A Numerical Model of Interactions between a Polar Ice Stream and the Ocean: Application to Ice Stream E, West Antarctica." *Journal of Geophysical Research: Oceans* 89(C3): 3523–3549.

Litfin, Karen T. 1994. *Ozone Discourses: Science and Politics in Global Environmental Cooperation*. New York: Columbia University Press.

Lomborg, Bjorn. 2001. *The Skeptical Environmentalist: Measuring the Real State of the World*. Cambridge: Cambridge University Press.

Longino, Helen E. 1990. *Science as Social Knowledge: Values and Objectivity in Scientific Inquiry*. Princeton, NJ: Princeton University Press.

Lucier, Paul. 2012. "The Origins of Pure and Applied Science in Gilded Age America." *Isis* 103(3): 527–536.

MacAyeal, D. R., and R. H. Thomas. 1982. "Numerical Modeling of Ice Shelf Motion." *Annals of Glaciology* 3: 189–193.

Mahlman, J. D., and S. B. Fels. 1986. "Antarctic Ozone Decreases: A Dynamical Cause?" *Geophysical Research Letters* 13: 1316–1319.

Mahoney, James R., John L. Malanchuck, and Patricia M. Irving. 1988. "Plan and Schedule for NAPAP's 1989 and 1990 Assessment Reports." *Journal of the Air Pollution Control Association* 38(12): 1489–1496.

Malone, E. L., and S. Rayner. 2001. "The Role of the Research Standpoint in Integrating Global-Scale and Local-Scale Research." *Climate Research* 19: 173–178.

Mann, Charles. 1990. "Meta-Analysis in the Breech." *Science* 249: 476–480.

Mann, Michael E. 2012. *The Hockey Stick and the Climate Wars: Dispatches from the Front Lines*. New York: Columbia University Press.

Markowitz, Gerald, and David Rosner. 2002. *Deceit and Denial: The Deadly Politics of Industrial Pollution*. Berkeley: University of California Press.

———. 2013. *Lead Wars: The Politics of Science and the Fate of America's Children*. Berkeley: University of California Press.

Martin, L. Robbin, Henry S. Judeikis, and Marilyn Wun. 1980. "Heterogeneous Reactions of Cl and ClO in the Stratosphere." *Journal of Geophysical Research* 85(C10): 5511–5518.

McCright, Aaron M., and Riley E. Dunlap. 2000. "Challenging Global Warming as a Social Problem." *Social Problems* 47(4): 499–522.

———. 2011. "Cool Dudes: The Denial of Climate Change among Conservative White Males in the United States." *Global Environmental Change* 21(4): 1163–1172.

McElroy, Michael B., Ross J. Salawitch, Steven C. Wofsy, and Jennifer A. Logan. 1986. "Reductions of Antarctic Ozone Due to Synergistic Interactions of Chlorine and Bromine." *Nature* 321: 759–762.

McLaughlin, S. B. 1985. "Effects of Air Pollution on Forests." *Journal of the Air Pollution Control Association* 35: 512–534.

McMillan, Thomas. 1988. Letter of Thomas McMillan to Lee Thomas, January 6. US Congress (House), 140–143.

Medvetz, Thomas. 2014. *Think Tanks in America*. Chicago: University of Chicago Press.

Mercer, J. H. 1978. "West Antarctic Ice Sheet and CO_2 Greenhouse Effect: A Threat of Disaster." *Nature* 271: 321–325.

Messer, Jay J. 2004. "Monitoring, Assessment, and Environmental Policy." In *Environmental Monitoring*, edited by G. Bruce Wiersma, 499–515. Boca Raton, FL: CRC.

Meyer, Judy L., Peter C. Frumhoff, Steven P. Hamburg, and Carlos de la Rosa. 2010. "Above the Din but in the Fray: Environmental Scientists as Effective Advocates." *Frontiers in Ecology and the Environment* 8(6): 299–305.

Michaels, David. 2008. *Doubt Is Their Product: How Industry's Assault on Science Threatens Your Health*. Oxford: Oxford University Press.

Miller, Clark A. 2004. "Climate Science and the Making of a Global Political Order." In *States of Knowledge*, edited by Sheila Jasanoff, 46–66. London: Routledge.

Miller, Clark A., and Paul N. Edwards. 2001. *Changing the Atmosphere: Expert Knowledge and Environmental Governance*. Cambridge, MA: MIT Press.

Mills, C. Wright. 2000. *The Power Elite*. New York: Oxford University Press.

Mitchell, R. B., W. C. Clark, D. W. Cash, and N. M. Dickson, eds. 2006. *Global Environmental Assessments: Information and Influence*. Cambridge, MA: MIT Press.

Mnookin, Seth. 2011. *The Panic Virus: A True Story of Medicine, Science, and Fear*. New York: Simon and Schuster.

Molina, L. T., M. J. Molina, R. A. Stachnik, and R. D. Tom. 1985. "An Upper Limit to the Rate of the Hydrogen Chloride + $ClONO_2$ Reaction." *Journal of Physical Chemistry* 89(18): 3779–3781.

Molina, M. J., and F. S. Rowland. 1974. "Stratospheric Sink for Chlorofluoromethanes: Chlorine Atom Catalysed Destruction of Ozone." *Nature* 249: 810–812.

Molina, M. J., T. Tso, L. T. Molina, F. C. Wang. 1987. "Antarctic Stratospheric Chemistry of Chlorine Nitrate, Hydrogen Chloride, and Ice: Release of Active Chlorine." *Science* 238: 1253–1257.

Mukerji, Chandra. 1989. *A Fragile Power: Scientists and the State*. Princeton, NJ: Princeton University Press.

Mumford, Lewis. 2010. *Technics and Civilization*. Chicago: University of Chicago Press.

Munk, Walter, and Roger Revelle. 1952. "On the Geophysical Interpretation of Irregularities in the Rotation of the Earth." *Geophysical Journal International* 6(6): 331–347.

———. 1952. "Sea Level and the Rotation of the Earth." *American Journal of Science* 250(11): 829–833.

NAPAP (National Acid Precipitation Assessment Program). 1983. *Annual Report to the President and Congress (1982)*. Washington, DC: Interagency Task Force on Acid Precipitation.

———. 1984. *Annual Report to the President and Congress (1983)*. Washington, DC: National Acid Precipitation Assessment Program.

———. 1987. *Annual Report to the President and Congress (1986)*. Washington, DC: National Acid Precipitation Assessment Program.

———. 1987. *NAPAP Interim Assessment*. 4 vols. Washington, DC: National Acid Precipitation Assessment Program.

———. 1988. *NAPAP Analysis of the Canadian RMCC Critique of the NAPAP Interim Assessment*. Washington, DC: National Acid Precipitation Assessment Program.

———. 1988. *NAPAP Analysis of the Critiques of NAPAP's 1987 Interim Assessment*. Washington, DC: National Acid Precipitation Assessment Program.

———. 1988. *Plan and Schedule for NAPAP Assessment Reports, 1989–1990. Public Review Draft*. Washington, DC: National Acid Precipitation Assessment Program.

———. 1989. *Models Planned for Use in the NAPAP Integrated Assessment*. Washington, DC: National Acid Precipitation Assessment Program.

———. 1991. *1990 Integrated Assessment Report*. Washington, DC: National Acid Precipitation Assessment Program.

———. 1991. *Acid Deposition: State of Science and Technology*. 4 vols. Edited by Patricia Irving. Washington, DC: National Acid Precipitation Assessment Program.

NAPAP and RMCC (Canadian Federal/Provincial Research Monitoring and Coordinating Committee). 1987. *Joint Report to the Bilateral Advisory and Consultative Group: The Status of Canadian/US Research in Acidic Deposition*. Washington, DC: National Acid Precipitation Assessment Program.

NASA (National Aeronautics and Space Administration), NOAA (National Oceanic and Atmospheric Administration), FAA (Federal Aviation Administration), WMO (World Meteorological Organization), and UNEP (United Nations Environment Programme). 1990. *Report of the International Ozone Trends Panel 1988*. 2 vols. Global Ozone Research and Monitoring Project Report no. 18. Geneva: WMO.

NASA, NOAA, NSF (National Science Foundation), and Chemical Manufacturers Association. 1987. "Antarctic Ozone: Initial Findings from Punta Arenas, Chile." September 30, 1987. http://cloud1.arc.nasa.gov/aaoe/project/statement.html (accessed January 28, 2017).

NASA, NOAA, UK DOE (Department of the Environment), UNEP, and WMO. 1992. *Scientific Assessment of Ozone Depletion: 1991*. Global Ozone Research and Monitoring Project Report no. 25. Geneva: WMO.

NASA, UK DOE, NOAA, UNEP, WMO. 1990. *Scientific Assessment of Stratospheric Ozone: 1989*. 2 vols. Global Ozone Research and Monitoring Project Report no. 20. Geneva: WMO.

NASA/WMO. 1982. *The Stratosphere 1981: Theory and Measurements*. WMO Global Ozone Research and Monitoring Project Report no. 11. Geneva: WMO.

———. 1986. *Atmospheric Ozone 1985: Assessment of Our Understanding of the Processes Controlling Its Present Distribution and Change*. 3 vols. WMO Global Ozone Research and Monitoring Project Report no. 16. Geneva: WMO.

National Research Council of Canada, Associate Committee on Scientific Criteria for Environmental Quality. 1981. *Acidification in the Canadian Aquatic Environment: Scientific Criteria for Assessing the Effects of Acidic Deposition on Aquatic Ecosystems*. Prepared by H. H. Harvey, R. C. Pierce, P. J. Dillon, J. R. Kramer, and D. M. Whelpdale. Ottawa: National Research Council of Canada.

Naylor, Simon, Katrina Dean, and Martin Siegert. 2008. "The IGY and the Ice Sheet: Surveying Antarctica." *Journal of Historical Geography* 34(4): 574–595.

NOAA, NASA, UNEP, and WMO. 1995. *Scientific Assessment of Ozone Depletion: 1994*. Global Ozone Research and Monitoring Project report no. 37. Geneva: WMO.

NOAA, NASA, UNEP, WMO, and EC (European Commission). 1999. *Scientific Assessment of Ozone Depletion: 1998*. Global Ozone Research and Monitoring Project Report no. 44. Geneva: WMO.

———. 2003. *Scientific Assessment of Ozone Depletion: 2002*. Global Ozone Research and Monitoring Project Report no. 47. Geneva: WMO.

———. 2007. *Scientific Assessment of Ozone Depletion: 2006*. Global Ozone Research and Monitoring Project Report no. 50. Geneva: WMO.

———. 2011. *Scientific Assessment of Ozone Depletion: 2010*. Global Ozone Research and Monitoring Project Report no. 52. Geneva: WMO.

———. 2014. *Scientific Assessment of Ozone Depletion: 2014*. Global Ozone Research and Monitoring Project Report no. 55. Geneva: WMO.

Norman, C. 1981. "Satellite Data Indicate Ozone Depletion." *Science* 213: 1088–1089.NRC. 1981. *Atmosphere-Biosphere Interactions: Toward a Better Understanding of the Ecological Consequences of Fossil Fuel Combustion*. Washington, DC: National Academy Press.

———. 1985. *Glaciers, Ice Sheets, and Sea Level: Effect of a CO_2-Induced Climatic Change*. Report of a Workshop Held in Seattle, Washington, September 13–15, 1984. Prepared by the Ad Hoc Committee on the Relationship Between Land Ice and Sea Level, Committee on Glaciology, Polar Research Board, Commission on Physical Sciences, Mathematics, and Resources, and National Research Council. Washington, DC: National Academy Press.

NRC, Committee on Atmospheric Transport and Chemical Transformation in Acid Precipitation. 1983. *Acid Deposition: Atmospheric Processes in Eastern North America: A Review of Current Scientific Understanding*. Prepared by Jack Calvert, James N. Galloway, Jeremy M. Hales, George M. Hidy, Jay Jacobson, Allan Lazrus, John Miller, Volker Mohnen, and Myron F. Uman. Washington, DC: National Academy Press.

NRC, Committee on the Institutional Means for Assessment of Risks to Public Health, Commission on Life Sciences. 1983. *Risk Assessment in the Federal Government: Managing the Process*. Washington, DC: National Academy Press.

NRC, US National Committee for CODATA (Committee on Data for Science and Technology Task Group on Chemical Kinetics), Commission on Physical Sciences, Mathematics and Applications. 1995. "The National Acid Precipitation Assessment Program." In *Finding the Forest in the Trees: The Challenge of Combining Diverse Environmental Data; Selected Case Studies*, 30–45. Washington, DC: National Academy Press.

Oden, Svante. 1967. "Nederbördens Försurning." *Dagens Nyheter*, October 24.

OECD (Organisation for Economic Co-operation and Development). 1977. *The OECD Programme on Long Range Transport of Air Pollutants: Summary Report*. Paris: Organisation for Economic Co-operation and Development.

Office of Science and Technology Policy. 1984. *Report of the Acid Rain Peer Review Panel*. Prepared by William A. Nierenberg and the Acid Rain Peer Review Panel. Washington, DC: Office of Science and Technology Policy.

Oppenheimer, Michael, Brian C. O'Neill, and Mort Webster. 2008. "Negative Learning." *Climate Change* 89: 155–172.

Oppenheimer, Michael, Brian C. O'Neill, Mort Webster, and Shardul Agrawala. 2007. "Climate Change: The Limits of Consensus." *Science* 317: 1505–1506.

O'Reilly, Jessica, Naomi Oreskes, and Michael Oppenheimer. 2012. "The Rapid Disintegration of Projections: The West Antarctic Ice Sheet and the Intergovernmental Panel on Climate Change." *Social Studies of Science* 42(5): 709–731.

Oreskes, Naomi. 1999. *The Rejection of Continental Drift: Theory and Method in American Earth Science*. New York: Oxford University Press.

———. 2010. "My Facts Are Better Than Your Facts: Spreading Good News about Global Warming." In *How Well Do Facts Travel?*, edited by Mary S. Morgan and Peter Howlett, 135–166. Cambridge: Cambridge University Press.

———. 2013. "The Scientist as Sentinel." *Limn* 3: 69–71.

———. 2015. "How Earth Science Became a Social Science." *Historical Social Research* 40(2): 246–270.

———. Forthcoming. *Science on a Mission: American Oceanography in the Cold War and Beyond*. Chicago: University of Chicago Press.

———. Forthcoming. *Should We Trust Science?* Princeton, NJ: Princeton University Press.

Oreskes, Naomi, and Kenneth Belitz. 2001. "Philosophical Issues in Model Assessment." In *Model Validation: Perspectives in Hydrological Science*, edited by M. G. Anderson and P. D. Bates, 23–41 London: John Wiley and Sons.

Oreskes, Naomi, and Erik M. Conway. 2010. *Merchants of Doubt: How a Handful of Scientists Obscured the Truth on Issues from Tobacco Smoke to Global Warming*. New York: Bloomsbury.

Oreskes, Naomi, Erik M. Conway, and Matthew Shindell. 2008. "From Chicken Little to Dr. Pangloss: William Nierenberg, Global Warming, and the Social Deconstruction of Scientific Knowledge." *Historical Studies in the Natural Sciences* 38(1): 109–152.

Oreskes, Naomi, Dale Jamieson, and Michael Oppenheimer. 2015. "What Role for Scientists?" In *The Proceedings of the Joint Workshop on Sustainable Humanity, Sustainable Nature, Our Responsibility, 2–6 May 2014*, edited by P. S. Dasgupta, V. Ramanathan, and M. Sanchez Sorondo, 617–649. Vatican City: Pontifical Academy of Sciences and Pontifical Academy of Social Sciences.

Oreskes, Naomi, and John Krige. 2014. *Science and Technology in the Global Cold War*. Cambridge, MA: MIT Press.

Oreskes, Naomi, and Ronald Rainger. 2000. "Science and Security before the Atomic Bomb: The Loyalty Case of Harald U. Sverdrup." *Studies in the History and Philosophy of Modern Physics* 31B: 309–369.

OTA (Office of Technology Assessment). 1984. *Acid Rain and Transported Air Pollutants: Implications for Public Policy*. Washington, DC: US Congress, Office of Technology Assessment.

Overrein, Lars N. 1976. "Report from the International Conference on the Effects of Acid Precipitation in Telemark, Norway, June 14–19, 1976." *Ambio* 5(5/6): 200–201.

Overrein, Lars N., Hans M. Seip, and Arne Tollan. 1980. *Acid Precipitation—Effects on Forests and Fish: Final Report of the SNSF Project 1972–1980*. Oslo: RECLAMO.

Oversight Review Board of the National Acid Precipitation Assessment Program. 1991. *The Experience and Legacy of NAPAP*. Washington, DC: Oversight Review Board of the National Acid Precipitation Assessment Program.

Paoletti, Elena, M. Schaub, R. Matyssek, G. Wieser, A. Augustaitis, A. M. Bastrup-Birk, A. Bytnerowicz, M. S. Günthardt-Goerg, G. Müller-Starck, and Y. Serengil. 2010. "Advances of Air Pollution Science: From Forest Decline to Multiple-Stress Effects on Forest Ecosystem Services." *Environmental Pollution* 158: 1986–1989.

Parson, Edward A. 2003. *Protecting the Ozone Layer: Science and Strategy*. New York: Oxford University Press.

———. "Stratospheric Ozone Collection, 1976–1997." Environmental Science and Public Policy Archives, Harvard University.

Patt, Anthony G. 1997. *Assessing Extreme Outcomes: The Strategic Treatment of Low Probability Impacts of Climate Change*. Environment and Natural Resources Program Discussion Paper E-97-10, Kennedy School of Government, Harvard University, August, International Institute for Applied Systems Analysis Interim Report IR-97-037, August.

———. 1999. "Extreme Outcomes: The Strategic Treatment of Low Probability Events in Scientific Assessments." *Risk Decision and Policy* 4(1): 1–15.

Pearce, Fred. 2010. *The Climate Files: The Battle for the Truth about Global Warming*. London: Guardian Books.

Perhac, Ralph. 1991. "Making Credible Science Usable: Lessons From CAA, NAPAP." *Power Engineering* 94: 38–40.

Peterson, Cass. 1987. "Downpour of Criticism Greets Acid-Rain Report." *Washington Post*, September 18.

Pielke, Roger A., Jr. 2007. *The Honest Broker: Making Sense of Science in Policy and Politics*. New York: Cambridge University Press.

Porter, Theodore M. 1995. *Trust in Numbers: The Pursuit of Objectivity in Science and Public Life*. Princeton, NJ: Princeton University Press.

———. 2010. *Karl Pearson: The Scientific Life in a Statistical Age*. Princeton, NJ: Princeton University Press.

Pratt, Mary T., and Pratt, Robert W. 1988. "Acid Rain: The Relationship between Sources and Receptors." In *Acid Rain: The Relationship between Sources and Receptors*, edited by James C. White, 1–8. New York: Elsevier.

Price, Derek J. de Solla. 1963. *Little Science, Big Science*. New York: Columbia University Press.

Proctor, Robert N. 1991. *Value-Free Science? Purity and Power in Modern Knowledge*. Cambridge, MA: Harvard University Press.

———. 2011. *Golden Holocaust: Origins of the Cigarette Catastrophe and the Case for Abolition*. Berkeley: University of California Press.

Proctor, Robert N., and Londa Schiebinger, eds. 2008. *Agnotology: The Making and Unmaking of Ignorance*. Stanford, CA: Stanford University Press.

Putnam, Hilary. 2002. *The Collapse of the Fact/Value Dichotomy and Other Essays*. Cambridge, MA: Harvard University Press.

Rahmstorf, Stefan. 2007. "A Semi-Empirical Approach to Projecting Future Sea-Level Rise." *Science* 315: 368–370.

Rainger, Ronald. 2001. "Constructing a Landscape for Postwar Science: Roger Revelle, the Scripps Institution, and the University of California, San Diego." *Minerva* 39(3): 327–352.

"The Report of the Royal Commission on Vaccination." 1896. *British Medical Journal* 2, no. 1864 (September 19): 526–528.

Revelle, Roger. 1982. "Carbon Dioxide and World Climate." *Scientific American* 247(2): 35.

Revelle, Roger, and Hans E. Suess. 1957. "Carbon Dioxide Exchange between Atmosphere and Ocean and the Question of an Increase of Atmospheric CO_2 during the Past Decades." *Tellus* 9(1): 18–27.

Rhodes, Richard. 1996. *Dark Sun: The Making of the Hydrogen Bomb*. New York: Simon and Schuster.

———. 2012. *The Making of the Atomic Bomb*. New York: Simon and Schuster.

Richardson, Katherine, and Ole Wæver. *2012*. "Building Bridges between Scientists and Policymakers to Reach Sustainability." *Solutions* 3(3): 107–110.

Rignot, E., J. L. Bamber, M. R. van den Broeke, C. Davis, Y. Li, W. J. van de Berg, and E. van Meijgaard. 2008. "Recent Antarctic Ice Mass Loss from Radar Interferometry and Regional Climate Modelling." *Nature Geoscience* 1: 106–110.

Rignot, E., G. Casassa, P. Gogineni, W. Krabill, A. Rivera, and R. Thomas. 2004. "Accelerated Ice Discharge from the Antarctic Peninsula Following the Collapse of Larsen B Ice Shelf." *Geophysical Research Letters* 31: L18401.

RMCC. 1986. *Assessment of the State of Knowledge on the Long-Range Transport of Air Pollutants and Acid Deposition: Part 1, Executive Summary*. Downsview, ON: Environment Canada.

———. 1987. *A Critique of the US National Acid Precipitation Assessment Program's Interim Assessment Report*. Downsview, ON: Federal/Provincial Research and Monitoring Coordinating Committee.

Roan, Sharon L. 1989. *Ozone Crisis: The 15-Year Evolution of a Sudden Global Emergency*. New York: John Wiley and Sons.

Roberts, Leslie. 1987. "Federal Report on Acid Rain Draws Criticism." *Science* 237: 1404–1406.

———. 1991. "Acid Rain Program: Mixed Review." *Science* 252: 371.

———. 1991. "Going for Broke on a Mega Model." *Science* 251: 1304.

———. 1991. "Learning from an Acid Rain Program." *Science* 251: 1302–1305.

Robock, Alan. 2010. "Nuclear Winter." *WIREs Climate Change* 1 (May/June): 418–427.

Rothschild, Rachel. 2014. "Burning Rain: The Long-Range Transboundary Air Pollution Project." In *Toxic Airs: Body, Place, Planet in Historical Perspective*, edited by James Rodger Fleming and Ann Johnson, 181–207. Pittsburgh: University of Pittsburgh Press.

Rowland, F. S. 1989. "Chlorofluorocarbons and the Depletion of Stratospheric Ozone." *American Scientist* 77: 36–45.

Rowland, F. S., H. Sato, H. Khwaja, and S. M. Elliott, 1986. "The Hydrolysis of Chlorine Nitrate and Its Possible Atmospheric Significance." *Journal of Physical Chemistry* 90: 1085–1088.

Royal Commission. 1898. *A Report on Vaccination and Its Results, Based on the Evidence Taken by the Royal Commission during the Years 1889–1897*. Vol. 1, *The Text of the Commission Report*. New Sydenham Society Publications, vol. 164. London: Adlard and Son.

Royal Commission on Environmental Pollution. 1974. *Pollution Control: Progress and Problems*. London: Her Majesty's Stationery Office.

"The Royal Commission on Vaccination." 1891. *Poor Law Magazine and Local Government Journal*.

Rubin, Edward S. 1991. "Benefit-Cost Implications of Acid Rain Controls: An Evaluation of NAPAP Integrated Assessment." *Journal of Air and Waste Management* 41: 914–921.

Rubin, Edward S., Lester B. Lave, and M. Granger Morgan. 1992. "Keeping Climate Research Relevant." *Issues in Science and Technology* 8(2): 47–55.

Sabin, Paul. 2013. *The Bet: Paul Ehrlich, Julian Simon, and Our Gamble over Earth's Future*. New Haven, CT: Yale University Press.

Sarewitz, Daniel. 2010. *Frontiers of Illusion: Science, Technology, and the Politics of Progress*. Philadelphia: Temple University Press.

Sato, H., and F. S. Rowland. 1984. "Current Issues in Our Understanding of the Stratosphere." Paper presented at the International Meeting on Current Issues in our Understanding of the Stratosphere and the Future of the Ozone Layer, Feldafing, West Germany.

Schefter, Jim. 1979. "New Aerodynamic Design, New Engines, Spawn a Revival of the SST." *Popular Science*, July, 62–65.

Schindler, David W. 1992. "A View of NAPAP from North of the Border." *Ecological Applications* 2: 124–130.

Schmandt, Jurgen, Judith Clarkson, and Hilliard Roderick. 1988. *Acid Rain and Friendly Neighbors: The Policy Dispute between Canada and the United States*. Durham, NC: Duke University Press.

Schneider, Keith. 1990. "Lawmakers Agree on Rules to Reduce Acid Rain Damage." *New York Times*, October 22.

Schneider, Stephen H. 2009. *Science as a Contact Sport: Inside the Battle to Save Earth's Climate*. Washington, DC: National Geographic Society Press.

Schofield, Carl L. 1976. "Acid Precipitation: Effects on Fish." *Ambio* 5(5/6): 228–230.

Schweber, Silvan S. 2007. *In the Shadow of the Bomb: Oppenheimer, Bethe, and the Moral Responsibility of the Scientist*. Princeton, NJ: Princeton University Press.

———. 2012. *Nuclear Forces: The Making of the Physicist Hans Bethe*. Cambridge, MA: Harvard University Press.

Shabecoff, Philipp. 1987. "Government Acid Rain Report Comes under Sharp Attack." *New York Times*, September 22.

Shackley, Simon, and Tora Skodvin. 1995. "IPCC Gazing and the Interpretive Social Sciences: A Comment on Sonja Boehmer-Christiansen's Global Climate Protection Policy: The Limits of Scientific Advice." *Global Environmental Change* 7: 77–79.

Sherwin, Martin. 1975. *A World Destroyed: The Atomic Bomb and the Grand Alliance*. New York: Vintage Books.

Smith, Alice Kimball. 1965. *A Peril and a Hope: The Scientists' Movement in America, 1945–47*. Chicago: University of Chicago Press.

Smith, Irene M. 1982. *Carbon Dioxide—Emissions and Effects*. Report no. IC TIS/TR 18. London: IEA Coal Research.

Smith, Robert A. 1852. "On the Air and Rain in Manchester." *Memoirs of the Literary and Philosophical Society of Manchester*, 2nd series, 10: 207–217.

Smithson, M. 1999. "Conflict Aversion: Preference for Ambiguity vs. Conflict in Sources and Evidence." *Organizational Behavior and Human Decision Processes* 79: 179–198.

Solomon, Susan. 2001. *The Coldest March: Scott's Fatal Antarctic Expedition*. New Haven, CT: Yale University Press.

Solomon, Susan, Rolando R. Garcia, F. Sherwood Rowland, and Donald J. Wuebbles. 1986. "On the Depletion of Antarctic Ozone." *Nature* 321: 755–758.

Stolarski, R. S., and R. J. Cicerone. 1974. "Stratospheric Chlorine: A Possible Sink for Ozone." *Canadian Journal of Chemistry* 52(8): 1610–1615.

Thomas, W. 2014. "Research Agendas in Climate Studies: The Case of West Antarctic Ice Sheet Research." *Climatic Change* 122(1–2): 299–311.

Thorpe, Charles. 2008. *Oppenheimer: The Tragic Intellect*. Chicago: University of Chicago Press.

Tung, K. K., M. W. K. Ko, J. M. Rodriguez, and N. D. Sze. 1986. "Are Antarctic Ozone Variations a Manifestation of Dynamics or Chemistry?" *Nature* 333: 811–814.

Uekoetter, Frank. 2009. *The Age of Smoke*. Pittsburgh: University of Pittsburgh Press.

UK DOE, Central Directorate on Environmental Pollution. 1979. *Chlorofluorocarbons and Their Effect On Stratospheric Ozone (Second Report)*. Pollution Paper no. 15. London: Her Majesty's Stationery Office.

UK DOE, Central Unit on Environmental Pollution. 1976. *Chlorofluorocarbons and Their Effect on Stratospheric Ozone*. Pollution Paper no. 5. London: Her Majesty's Stationery Office.

Urofsky, Melvin I. 2015. *Dissent and the Supreme Court: Its Role in the Court's History and the Nation's Constitutional Dialogue*. New York: Pantheon Books.

US Congress. 1979. *Congressional Record*. 96th Cong., 1st sess., vol. 125, pt. 19.

US Congress (House), Subcommittee on Energy Development and Applications and Subcommittee on Natural Resources, Agriculture Research, and Environment of the Committee on Science and Technology. 1983. *Acid Rain: Implications for Fossil R&D*. 98th Cong., 1st Sess., September 13 and 20.

US Congress (House), Subcommittee on Natural Resources, Agriculture Research, and Environment of the Committee on Science and Technology. 1981. *Acid Rain*. 97th Cong., 1st Sess., September 18 and 19, November 19, and December 9.

———. 1988. *National Acid Precipitation Assessment Program*. 100th Cong., 2nd Sess., April 27.

US Congress (Senate), Committee on Environment and Public Works. 1985. *Review of the Federal Government's Research Program on the Causes and Effects of Acid Rain*. 99th Cong., 1st Sess., December 11.

US Department of Energy. 1979. *Multistate Atmospheric Power Production Pollution Study—MAP3S: Progress Report for FY 1977 and FY 1978*. Prepared by Michael C. MacCracken et al. Washington, DC: US Department of Energy.

US Department of Transportation, Climate Impacts Assessment Program. 1975. *I: Panel on the Natural Stratosphere*. DOT-TST-75-51. Washington, DC: Institute for Defense Analysis.

———. 1975. *II: Propulsion Effluents in the Stratosphere*. DOT-TST-75-52. Washington, DC: Institute for Defense Analysis.

———. 1975. *III: The Stratosphere Perturbed by Propulsion Effluents*. DOT-TST-75-53. Washington, DC: Institute for Defense Analysis.

———. 1975. *IV: The Natural and Radiatively Perturbed Troposphere*. DOT-TST-75-54. Washington, DC: Institute for Defense Analysis.

———. 1975. *V: Impacts of Climatic Change on the Biosphere*. DOT-TST-75-55. Washington, DC: Institute for Defense Analysis.

US EPA (Environmental Protection Agency). 1983. *Acid Deposition Phenomenon and Its Effects: Critical Assessment Review Papers*. 2 vols. Washington, DC: US EPA.

———. 1983. *Projecting Future Sea Level Rise: Methodology, Estimates to the Year 2100, and Research Needs*. Edited by John S. Hoffman, Dale Keyes, and James G. Titus. Washington, DC: EPA.

US General Accounting Office. 1987. *Acid Rain: Delays and Management Changes in the Federal Research Program*. Washington, DC: US General Accounting Office.

———. 1989. *Air Pollution: Improved Atmospheric Model Should Help Focus Acid Rain Debate*. Washington, DC: US General Accounting Office.

US NAS (National Academy of Sciences), Climate Impact Committee. 1975. *Environmental Effects of Stratospheric Flight: Biological and Climatic Effects of Aircraft Emissions in the Stratosphere*. Washington, DC: National Academy of Sciences.

US NAS, Committee on Impacts of Stratospheric Change. 1976. *Halocarbons: Environmental Effects of Chlorofluoromethane Release*. Washington, DC: National Academy of Sciences.

US NAS, Panel on Atmospheric Chemistry. 1975. *Atmospheric Chemistry: Problems and Scope*. Washington, DC: National Academy of Sciences.

———. 1976. *Halocarbons: Effects on Stratospheric Ozone*. Washington, DC: National Academy of Sciences.

The Vaccination Inquirer and Health Review. 1882. London: London Society for the Abolition of Compulsory Vaccination.

Vardy, Mark, Michael Oppenheimer, Navroz Dubash, Jessica O'Reilly, and Dale Jamieson. 2017. "The Intergovernmental Panel on Climate Change: Challenges and Opportunities." *Annual Review of Environment and Resources* 42: 55–75.

Vasileiadou, Eleftheria, Gaston Heimeriks, and Arthur C. Petersen. 2011. "Exploring the Impact of the IPCC Assessment Reports on Science." *Environmental Science and Policy* 14: 1052–1061.

Vaughan, David G., and John R. Spouge. 2002. "Risk Estimation of Collapse of the West Antarctic Ice Sheet." *Climatic Change* 52(1): 65–91.

Vaughan, Diane. 1996. *The Challenger Launch Decision*. Chicago: University of Chicago Press.

Velicogna, I. 2009. "Increasing Rates of Ice Mass Loss from the Greenland and Antarctic Ice Sheets Revealed by GRACE." *Geophysical Research Letters* 36: L19503.

Vermeer, M., and S. Rahmstorf. 2009. "Global Sea Level Linked to Global Temperature." *Proceedings of the National Academy of Sciences* 106: 21527–21532.

Wang, Jessica. 1999. *American Science in an Age of Anxiety: Scientists, Anticommunism, and the Cold War*. Chapel Hill: University of North Carolina Press.

Wang, Zuoyue. 1997. "Responding to Silent Spring: Scientists, Popular Science Communication, and Environmental Policy in the Kennedy Years." *Science Communication* 19: 141–163.

———. 2008. *In Sputnik's Shadow: The President's Science Advisory Committee and Cold War America*. New Brunswick, NJ: Rutgers University Press.

———. 2014. "Scientists and Arms Control: The US-China Case and Comparisons with Climate Change." Paper presented at the Arms Control and Climate Change Conference, University of Texas at Austin, January 16–17, 2014.

Weart, Spencer. 1979. *Scientists in Power*. Cambridge, MA: Harvard University Press.

———. 2003. *The Discovery of Global Warming*. Cambridge, MA: Harvard University Press.

Weertman, J. 1974. "Stability of the Junction of an Ice Sheet and an Ice Shelf." *Journal of Glaciology* 13: 3–11.

Weinberg, Alvin M. 1968. *Reflections on Big Science*. Cambridge, MA: MIT Press.

Weingart, Peter. 2002. "The Moment of Truth for Science." *EMBO Reports* 3(8): 701–803.

Wetstone, Gregory S., and Armin Rosencranz. 1984. *Acid Rain in Europe and North America: National Responses to an International Problem*. Washington, DC: Environmental Law Institute.

White, James C., ed. 1988. *Acid Rain: The Relationship between Sources and Receptors*. New York: Elsevier.

Whyte, William H. 2013. *The Organization Man*. Philadelphia: University of Pennsylvania Press.

Winter, Alison. 1998. *Mesmerized: Powers of Mind in Victorian Britain*. Chicago: University of Chicago Press.

Wolfe, Robert M., and Lisa K. Sharp. 2002. "Anti-Vaccinationists Past and Present." *British Medical Journal* 325(7361): 430–432.

Wormbs, Nina, and Sverker Sörlin. 2017. "Arctic Futures: Agency and Assessing Assessments." In *Arctic Environmental Modernities: From the Age of Polar Exploration to the Era of the Anthropocene*, edited by L. A. Körber, S. MacKenzie, and A. Westerståhl Stenport, 247–261. Palgrave Studies in World Environmental History. Cham, Switzerland: Palgrave Macmillan.

Wuebbles, D. J. 1983. "Chlorocarbon Emission Scenarios: Potential Impact on Stratospheric Ozone." *Journal of Geophysical Research* 88: 1433–1443.

Yearley, S. 2009. "Sociology and Climate Change after Kyoto: What Roles for Social Science in Understanding Climate Change?" *Current Sociology* 57(30): 389–405.

York, Herbert. 1989. *The Advisors: Oppenheimer, Teller, and the Superbomb*. Stanford, CA: Stanford University Press.

Zalasiewicz, Jan, et al. 2015. "When Did the Anthropocene Begin? A Mid-Twentieth Century Boundary Level Is Stratigraphically Optimal." *Quaternary International* 383: 196–203.

Zamitto, John H. 2004. *A Nice Derangement of Epistemes: Post-Positivism in the Study of Science from Quine to Latour*. Chicago: University of Chicago Press.

Index